20 Reasons to Question Plate Tectonics

© 2023 by Ellis Hughes
Published and printed in the USA. All rights reserved.
THM Books
ISBN 978-1-7374763-9-9
questionplatetectonics@outlook.com

Bulk or retail orders: ingramcontentgroup.com

Please note that the inclusion of data, ideas, opinions, and quotes by various researchers and authors does not imply their endorsement of this book.

TABLE OF CONTENTS

INTRODUCTION

A BRIEF HISTORY OF PLATE TECTONIC THEORY

Plate tectonics is a theory that proposes that the earth's lithosphere (the crust and upper mantle) is broken into large sections, or "plates," and that the movement of these plates is the cause of earth's major topological features including mountains, valleys, and ocean trenches. The force that moves the plates is believed to be upwellings of magma from deep within the mantle, creating new crust at divergent zones (mainly new seafloor at mid-ocean ridges), and pushing older crust back down into the mantle at convergent (subduction) zones. These ideas will be discussed in more depth after we learn how and when this theory came about.

The history of the theory

Until the 1900s, most geologists assumed that earth's bumps and wrinkles had been formed either as the result of gradual shrinking, or because of the vertical motion of the continents as they went through cycles of rising and sinking. No one in "mainstream" geology thought that the continents were moving over the earth's surface.

The earliest record of a geologist suggesting that land masses moved laterally was a man named Abraham Ortelius in the year 1596. World maps had become accurate enough for Ortelius to observe the similarity between the coastlines of North and South America and the coastlines of Europe and Africa. However, no one at that time took any notice of this idea, and it quickly faded away. The next person to suggest that continents could slide was Roberto Mantovani in 1889. He had studied the geology of both Africa and South America. He noticed that the rocks on the edges of these continents were very similar and he wondered if the continents were once much closer together, if not actually joined. Two decades later, in 1908, a geologist named Frank Taylor proposed that the continents were "creeping" toward the equator due to earth's recent capture of the moon during the dinosaur era. Taylor also thought that the Himalayas were the result of a collision between India and Asia. A few other scientists were saying similar things, but their writings were never widely circulated.

One man has gone down in history as the author of the theory of continental movement: a meteorologist named Alfred Wegener, who pioneered the use of weather balloons, and went on expeditions to Greenland to study polar weather. By 1912, Wegener had read the papers written by Mantovani, Taylor, and others, and put all the ideas together to create a theory he called

Wegener in Greenland, winter of 1912-13

"continental drift." He believed that the continents had started out as one large land mass that he named Pangea. He said that Pangea had somehow broken apart and the pieces had started drifting in different directions at a speed of about 250 centimeters (100 inches) a year. He had some very good evidence that the continents had been joined together, such as similar fossils on both sides of the Atlantic Ocean. However, Wegener's proposal did not include a mechanism that could explain how the continents broke apart, nor

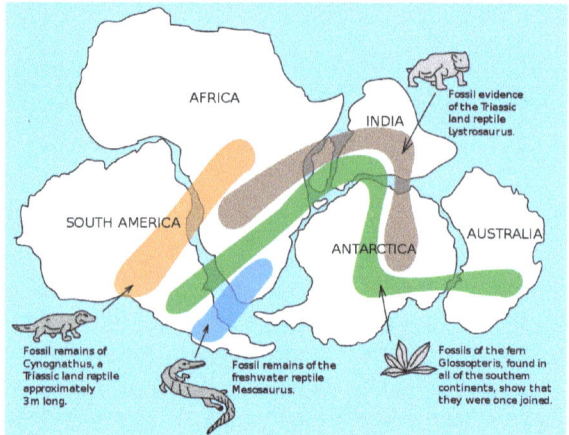

AFRICA
INDIA
Fossil evidence of the Triassic land reptile *Lystrosaurus*.
SOUTH AMERICA
ANTARCTICA
AUSTRALIA
Fossil remains of *Cynognathus*, a Triassic land reptile approximately 3m long.
Fossil remains of the freshwater reptile *Mesosaurus*.
Fossils of the fern *Glossopteris*, found in all of the southern continents, show that they were once joined.

This public domain USGS image shows up frequently in articles about Plate Tectonic Theory. The image continues to be used even though some of these fossils have been found in other parts of the world, weaking the strength of the argument.

any suggestions as to what force could have caused their movement. Because of these two issues, and the fact that Wegener did not have a degree in geology, his theory was not accepted by the major geological societies of his day. In fact, his theory was very unpopular for almost 40 years. In the 1920s, the American Association of Petroleum Geologists organized a symposium specifically to oppose the hypothesis of continental drift. They agreed with many European geologists that oceanic crust was too firm for continents just to "plow through," and that Wegener's estimate of continental move-ment at the rate of 250 centimeters a year was unreasonable.

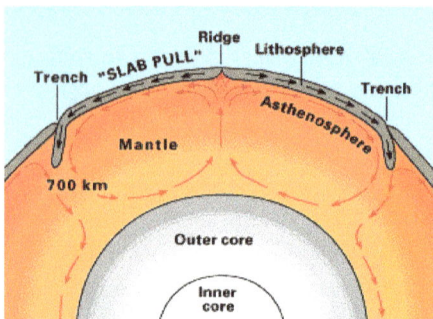

Ridge
Trench "SLAB PULL"
Lithosphere
Trench
Asthenosphere
Mantle
700 km
Outer core
Inner core

In 1919, British geologist Arthur Holmes, a supporter of Wegener's ideas, proposed a mechanism that might be able to drive continental drift. He suggest-ed that the earth's mantle was heated by radioactivity and that it circulated in convection cells, much like boiling water circulates in a pot. This circulation inside the mantle, he speculated, caused the crust above it to shift and slide. In his life's work, *Principles of Physical Geology*, published in 1944, Holmes devoted an entire chapter to this topic.

In the 1920s and 1930s, Dutch geophysicist Felix Andries Vening Meinesz made numerous voyages deep into the ocean using submarines. He was especially interested in earth's magnetism, and he invented a way to measure gravity. He found that the gravity over the Puerto Rican Trench was surprisingly low. After analyzing the data from many of his ocean explorations, he became a supporter of the idea of continen-tal movement, and thought that ocean trenches might be the result of earth's crust having been pushed together at these places.

In the 1930s, American geophysicist William Maurice Ewing picked up where Meinesz left off, and during the next few decades took the study of gravity, magnetism, and seismology to a much higher level. Ewing had it all: an excellent knowledge of physics, practical skills needed to improve and build tools, leadership skills, attention to detail (such as data collection), a healthy pair of "sea legs," and a sense of adventure. His first expedition used seismic data to explore the continental shelf off the coast of Virginia. Not much was known about these "shelves" that extended for many miles then dropped off suddenly into

Maurice Ewing at work, 1948.

deep ocean. Were these shelves merely rubbish that had accumulated at the edges of the continents? Was there a deep fault line where they dropped off? Were they submerged continents from bygone eras? Ewing discovered a pile of sediments 3,800 meters thick, and underneath this layer there appeared to be large reservoirs of oil. Ewing approached an oil company in 1936 and suggested that there might be profit in drilling for oil off shore, on these shelves. The oil executive told him that there was plenty of oil on land and his company was not in the least bit interested in going out to sea to look for it.

Ewing then worked with Harry Hess (who plays an important role later) to continue the study of gravity in various parts of the ocean. They found an arc-shaped zone of low gravity (a "gravity deficit") in the Caribbean around the Lesser Antilles. Meinesz had already found low gravity arcs around Puerto Rico and in Indonesia.

Always looking for a new frontier, Ewing extended his work farther out into the Atlantic. He found large ripple marks on the seafloor in the Gulf of Maine. This was the first of his many discoveries that would surprise and puzzle geologists.

During World War II, Ewing worked for the Woods Hole Oceanographic Institute and discovered the SOFAR (Sound Fixing and Ranging) channel. At a certain depth, low frequency sound waves are able to travel long distances without dissipating. The Navy began using this channel for communications between submarines. After the war, Ewing took a post at Columbia University, and they supplied him with a research team and a vessel that allowed him to make the additional discoveries that led up to the formation of Plate Tectonic theory.

In the late 1940s, Ewing discovered the Mohorovičić discontinuity under the ocean.

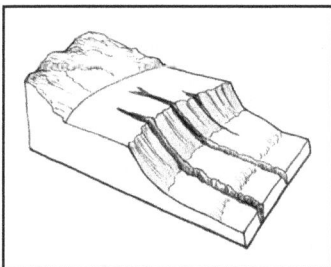

The "Moho" is assumed to be the dividing line between the crust and the mantle. Under continents, it is 20 to 60 km deep. Ewing found that under the ocean it can be as shallow as 5 km. His seismic data also suggested that oceanic crust is made of basalt, not granite like continental crust.

Ewing and his devoted team then began exploring the edges of the continental shelf and the abyssal plains beyond them. They were stunned

3

when they found massive canyons cut into the edges of the continental shelf and continuing out into the plains for hundreds of miles. These canyons could not have been carved underwater. Some of them had been cut into very hard rock. Geologists were (and still are) mystified.

In the 1950s, Ewing turned his attention to the Mid-Atlantic Ridge. This underwater mountain range had been discovered in the late 1800s by the ships that had laid the telegraph cables across the ocean. The MAR had already been explored to some degree in the early 1900s, but Ewing made a new discovery about the ridge: it had a deep valley running down the center. He also documented many minor earthquakes coming from the ridge and guessed that the ridge could be traced around the world simply by using earthquake data. He took many core samples from the ridge and found that no two were exactly alike. After examining all his data, Ewing came to the conclusion that the ocean, in general, and the Mid-Ocean Ridges in particular, were far younger than the continents.

Finally, in 1952, Ewing made his last major contribution. He took an improved version of the magnetometer (invented by Meinesz) and towed it behind his ship. He sailed back and forth in the North Atlantic and recorded data. The result was a striped pattern going east to west. Ewing did not interpret these stripes, he just recorded them. In 1959, he worked alongside Marie Tharp and Bruce Heezen to published these findings in a book titled, *The Floors of the Oceans; The North Atlantic*.

Once this data became available to all geologists, theories and opinions began to form. In 1960, Bruce Heezen wrote a paper suggesting that the mid-ocean ridges were places where new seafloor was forming. This would explain the data suggesting that oceanic crust was younger than continental crust. This paper was immediately followed by papers written by Harry Hess and Robert Dietz, who laid out a theory that incorporated some previous ideas along with the recent data.

The theory proposed by Hess and Dietz

Hess and Dietz both proposed a scenario in which new seafloor was being formed at the mid-ocean ridges. They guessed that perhaps the ridges were a weak point in the earth's crust. Magma was rising from deep in the mantle and pushing its way out along the crack in the middle of the ridge. Over time, this newly formed crust on the ridge would be displaced by fresh magma and therefore be placed by fresh magma and therefore be pushed to the sides. This became known as "seafloor spreading." Over millions of years, that new piece of seafloor wouldn't be new anymore and it would slowly move along, as if on a conveyor belt, and would eventually find itself at a place where it would be pushed back down into the mantle again. These areas where seafloor was being pushed back into the mantle became known as "subduction zones." The deep ocean trenches in the western Pacific were believed to be subduction zones. Also, many places along the rim of the Pacific Ocean (the so-called "Ring of Fire") would be sites of subduction.

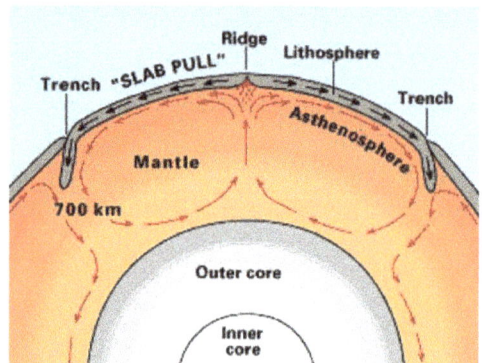

Magnetic striping added to the theory

In 1963, Lawrence Morley, Fred Vine and Drummond Matthews published their hypothesis that as lava comes out at the mid-ocean ridges, it cools in a way that reflects the magnetic properties of the earth at that time. Magma made from basalt is high in iron, often in the form of the mineral magnetite, Fe_3O_4. At high temperatures, the magnetic properties of magnetite disappear. The exact point at which this happens is called the Curie point, after the discoverer of this phenomenon, Pierre Curie. As magma cools and goes below its Curie point, the magnetic properties of its magnetite return. Before it solidifies, magnetite is affected by the earth's strong magnetic field, and its own magnetic field will align with that of the earth. Once cold and hard, the magnetite's magnetic orientation is "locked in" and becomes permanent. Oceanic crust now came to be seen as a "tape recorder" of past events, with the stripes representing magnetic reversals. Going in both directions out from the Mid-Ocean Ridge, the stripes took you farther back in time.

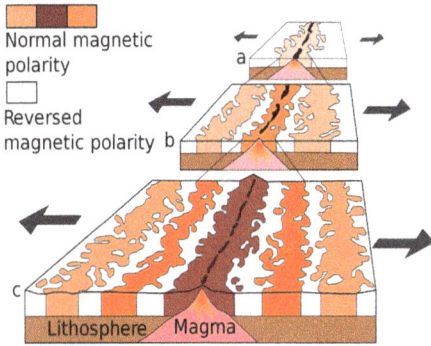

General acceptance of the theory

In 1964, Tuzo Wilson, a strong supporter of continental drift and sea floor spreading, added a new idea to the theory: transform faults that could slide back and forth. This allowed for additional explanations of how and why plates could move. In the same year, Edward Bullard used a computer simulation (a novel technique in that day!) to show how the continents could have been connected along the Mid-Atlantic Ridge. This has become known as the "Bullard's Fit" of the continents.

At the 1967 meeting of the American Geophysical Union, Plate Tectonic Theory was generally accepted by all those present. The following year, Xavier Le Pichon wrote a paper describing a global scheme of plate motion, and this was added to the theory. Also around this time, the theory began to incorporate research done that had been done by Hugo Benioff in the 1940s and 50s. Benioff found planes of earthquake activity near trenches, some of which seemed to go down at a 45 degree angle.

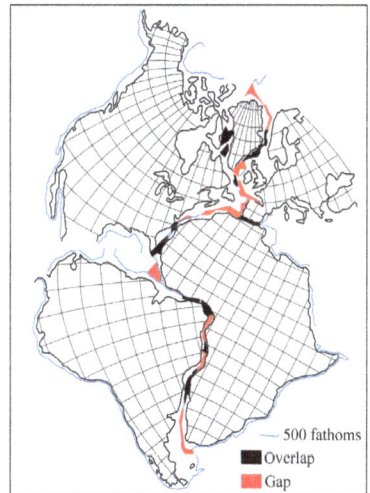

The "Bullard's Fit"

Jacques Kornprobst (redesigned after Bullard, E., Everett, J.E. and Smith, A.G., 1965. The fit of the continents around the Atlantic. Phil. Trans. Royal Soc., A 258, 1088, 41-51), CC BY-SA 4.0 <https://creativecommons.org/licenses/by-sa/4.0>, via Wikimedia Commons

Summary of Plate Tectonic Theory as it is held today

1) Earth's lithosphere is composed of seven or eight major plates (depending on how they are defined) and many minor plates. Where the plates meet, their relative motion determines the type of boundary: convergent, divergent, or transform. Divergent boundaries are where plates are moving apart and new seafloor is being created. Convergent boundaries are places where one plate is diving under, or going over, another. Transform boundaries are places where plates are moving back and forth laterally.

2) Plate tectonic processes began 3.2 to 3.5 billion years ago.

3) Plate boundaries are associated with earthquakes, volcanic activity, formation of ocean trenches, and mountain-building.

4) The speed at which plates move can range from zero to 10 cm per year.

5) Continental plates are made of either continental or oceanic lithosphere topped by the appropriate type of crust (continental or oceanic). Plates are able to move because lithospheric rock has greater mechanical strength than the underlying asthenosphere.

6) Along convergent boundaries, one plate dives under another through a process called subduction. The diving plate sinks down into the mantle, and will eventually be recycled as it becomes part of the mantle. Most convergent boundaries lie at the edges of oceans or at ocean trench zones. However, mountainous areas may also be convergent zones, with the mountains being places where plates have "piled up" instead of disappearing back into the mantle.

7) The sea floor that is lost in the process of subduction is offset by the creation of new sea floor at divergent zones. This means that the youngest crust lies along the mid-ocean ridges and it gets older the further away you go from the ridges.

8) Fracture zones along the mid-ocean ridges are places where uneven spreading pressure has been released, so these fracture zones must point in the direction of seafloor spreading.

9) The driving force that moves tectonic plates is convection of the mantle. The mantle has various convection cells. The rising portion of a cell is thought to be where a plume of hot material from the outer core is rising through the mantle.

10) Benioff zones (also known as Wadati-Benioff zones) are angular, planar earthquake zones at convergent zones and are believed to show places where one plate is diving under another.

11) As plates move, sometimes they go over "hotspots" in the mantle. This results in a chain of volcanoes (e.g. Hawaii).

EARLY OBJECTIONS TO THE THEORY

Not everyone was convinced that the plate tectonic (PT) hypothesis was the best solution to the continental puzzle. Early dissenters included German-American geologist Curt Teichert, American geologist Walter H. Bucher, Australian geologist Rhodes Fairbridge, Dutch geologist D. Van Hilten, Russian scientists Vladimir Beloussov and E. N. Lyustikh, British mathematician and geophysicist Sir Harold Jeffreys, and American geologists Howard and Arthur Meyerhoff (father and son). These men were well-respected in their fields, and some had received prizes and medals from universities or scientific societies. They were as qualified to write about geology and geophysics as the early PT theorists were.

In the same year that Plate Tectonic theory was officially adopted by the GSA, E. N. Lyustikh of the Institute of Physics of the Earth, Moscow, wrote a succinct, 5-page paper giving a number of objections, but focusing primarily on the problem of convection in the mantle and the faulty application of Rayleigh's work on spherical harmonics to the proposed "convection cells" in the mantle. Lyustikh also pointed out the pitfall of relying too heavily on theoretical mathematical models instead of using real-life data. He saw many hidden assumptions baked into the PT explanations, such as the assumption that the surface of the asthenosphere is a perfect sphere (when it is almost certainly not). Lyustikh was not at all impressed with the Bullard's fit of the continents and he showed how easy it is to find matching coastlines all over the world. He criticized Robert Dietz's idea of seafloor spreading by saying that according to the principles of geophysics, if there is a rising heat current under the mid-ocean ridges, we should find a negative gravity anomaly there, and none has been found. He ended by stating that he agreed with the opinion of Sir Harold Jeffreys, that "the articles of the drifters [PT advocates] are remarkable for fallacious data, misinterpretation of the data, and omission to mention any objections."[1]

In 1970, V. V. Beloussov published a paper titled, "Against the Hypothesis of Ocean-floor Spreading."[2] After pointing out problems associated with rocks found on the Mid-Atlantic Ridge, he wrote extensively about issues with magnetic anomalies (magnetic striping) associated with mid-ocean ridges. Since the most cited example of magnetic striping is right below Iceland, Beloussov pointed out that if you follow these magnetic stripes north, right into Iceland itself, you have the perfect place to test your theory—the rocks are right there in front of you. The magnetic data should match the geological data. It turns out, he says, that there is a disconnect between the assumed ages of the magnetic stripes and the estimated ages of the visible rocks. Beloussov pointed out a few other details about plate tectonics that don't make sense, such as the immense volume of rock that is supposed to be disappearing into the mantle at subduction zones, and the belief that the cooling of the lithosphere is what causes it to sink, despite the observation that the ocean floor isn't any cooler at the edges than it is only a few miles from the ridge.

In 1972, Howard and Arthur Meyerhoff published a paper in the *Journal of Geology*, in which they introduced an alternative to Plate Tectonics: a model they called the Fracture-Contraction Hypothesis.[3] The Meyerhoffs' work was based on an earlier model known as the Davison-Darwin-Jeffreys model. (Sir G. H. Darwin, not Charles Darwin.) In the opening paragraphs of their paper, the Meyerhoffs accuse the "new global tectonics model," as PT was known then, of being a "Procrustean bed." In Greek mythology, Procrustus was a sly and wicked innkeeper who would lure travelers in for the night and make them sleep in an iron bed. If they were too short for the bed he would stretch them to make them fit. If they were too long, he would lop off body parts until they fit. The Meyerhoffs apparently saw data being either stretched, or ignored.

By the late 1970s, groupthink had taken over, and the editorial boards of most professional journals were pressured by their financial sponsors not to publish papers that argued against plate tectonics.[4] Grant money was denied to anyone suspected of not being on board with the PT program, and since this type of research takes large ships and expensive equipment, dissenters were effectively squelched. This was before the age of the Internet and self-publishing, so the refusal of journals to accept articles with alternative viewpoints led to the banishment of those ideas from mainstream science. Fortunately, a few minor journals did publish some papers, and the papers are now part of their digital archives. This gives us a paper trail, albeit a narrow one, documenting the early history of dissent.

These early objectors had protégés that took up the cause and formed a second generation of dissenters who continued the struggle against what they believed to be a deeply flawed theory. Some of these scientists will be mentioned in the last section of this book, where alternative theories are discussed.

REASONS TO QUESTION PLATE TECTONICS

REASON #1: There are too many miles of divergent boundaries.

As soon as the idea of seafloor spreading was conceived, an obvious question arose: Where is all the old seafloor going? If the earth is constantly creating new crust and has no way to dispose of old crust, then the earth should be expanding. Scientists have looked at historical data about the size of the earth and have determined that the earth is not expanding; the size appears to be quite stable. Therefore, to keep the size of the lithosphere constant, there must be an approximately equal number of miles of divergence and convergence. For every square mile of new crust being formed at divergent zones, there must be a square mile of crust that somehow disappears at convergent zones.

According to PT, divergent boundaries occur mainly at the Mid-Ocean Ridges. We can trace the ridges around the globe to calculate total mileage: about 74,000 km. If we assume that spreading is occurring on both sides of the ridge, that's a total of 148,000 km of divergence. We need to come up with an equal amount of convergence. Ocean trenches are convergent zones and they give us about 30,500 km. Collision zones where mountains are forming are also classified as convergent, and they give us another 11,000 km. (Mediterranean-Zagros-Himalayan-Indonesian zone: 9,000 km, Banda Arc in Indonesia: 2,000 km.) If we include smaller zones in various places, we can bring the tally up to 55,000—the number quoted in a (pro-PT) article in "Reviews of Geophysics" (2002).[5]

We have 148,000 km of divergent boundaries and only 55,000 km of convergent. We can see this inequality on a map. The green lines are divergent zones (spreading centers). The red lines are convergent zones, where sea floor should be disappearing. Purple lines are "transform faults" and are neither divergent nor convergent. (Note: Maps like this can vary. You might find one that is slightly different, but the differences will be relatively small.)

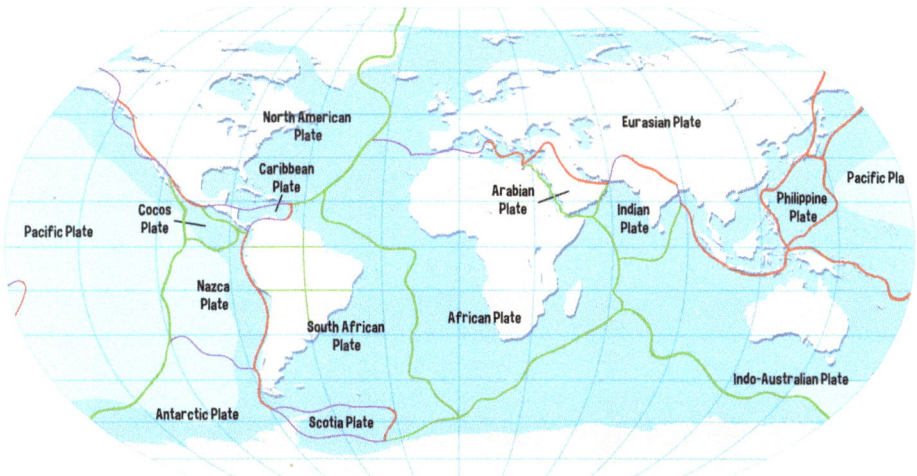

Let's look at a map with arrows showing what is supposed to be happening at each boundary. Arrows pointing in opposite directions indicate spreading; arrows pointing at each other indicate convergence (subduction or collision). This map is probably the most widely used plate tectonic map in the world, since it appears in the Wikipedia article on plate tectonics.

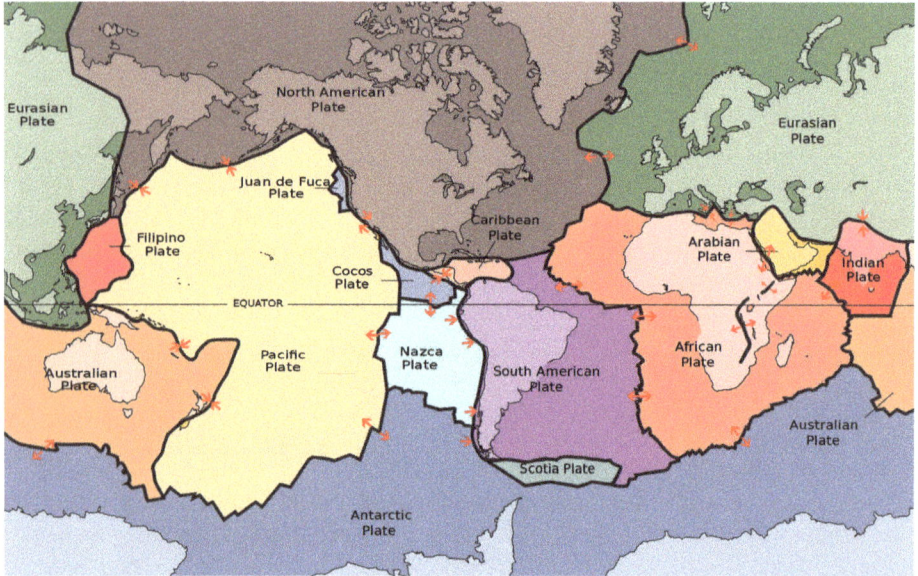

Look at the line that goes across the southern edge of all the plates (the bottom of the Australian, Pacific, Nazca, South American, and African plates). We are not overly supplied with arrows along this line, but the few we are given definitely seem to indicate that most of this line is a divergent boundary where seafloor is spreading. Now look at Antarctica. We don't see even a hint of a convergent boundary around it. Where does all that southbound seafloor go? Early critics saw this as a fatal flaw that should have canned the theory immediately.

Look at the eastern edge of the Nazca plate. It is thought to be disappearing at the convergent zone that runs along the western side of South America. This is what the theory predicts and expects. Now look at the western and southern borders of the African plate. That's a lot of divergent mileage. We don't see a convergent zone along the west coast of Africa, like we do along the west coast of South America. The nearest convergent boundary for Africa is the tiny Aegean trench zone in the Mediterranean Sea. Are we expected to believe that the entire African plate will eventually disappear into this tiny trench? Also notice the divergent zone that runs north/south inside Africa, along the famous "rift zone." Where will all that new crust go? Some theorists predict that the rift will eventually open so wide that a new ocean will form. How will that happen without any convergent boundaries providing the slab pull force necessary for plate motion? Europe has the same problem. Notice the amazing lack of convergent zones all along its western edge. Is the spreading at the Mid-Atlantic Ridge able to push the entire Asian plate?

One can't help but think that the original theorists really expected to find evidence of subduction along the Atlantic continental shelves of North and South America. If subsequent ocean floor research had found trenches along these coastlines, this would have provided much-needed validation for the theory. Instead, they found massive submarine canyons that could only have been carved if the shelves had been much higher than sea level at some point in the past. There are places in the world where we can watch underwater erosion happening and see what it does to the seafloor landscape. It doesn't carve canyons like these.

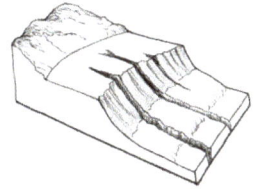

Some boundary lines are classified as neither divergent nor convergent and therefore did not contribute to our total mileage calculations. These "transform" boundaries are thought to be places where the plates slide sideways. One of the largest transform boundaries deserves special mention: the boundary between the North and South American plates.

The original theorists didn't like the idea of having both North and South America on the same plate. A combined plate would cover half the globe and create issues that would probably be fatal to the theory. So they just assumed a boundary had to be there somewhere, and drew a line on the map. Later, when more detailed maps of the Atlantic ocean floor became available, the boundary line was drawn along one of the many east/west fracture zones (east-west lines in the diagram) that cross the Mid-Ocean Ridge. The chosen fracture zone looks exactly like all the other fracture zones. (Can't see it inside the red circle? Neither can anyone else.) Scientific proof that this particular fracture zone is, in fact, the boundary line, is lacking, so information about this boundary is hard to find. Articles written about transform boundaries between plates focus almost entirely on well-established sites, such as the San Andreas fault zone, and rarely mention this important Atlantic boundary.

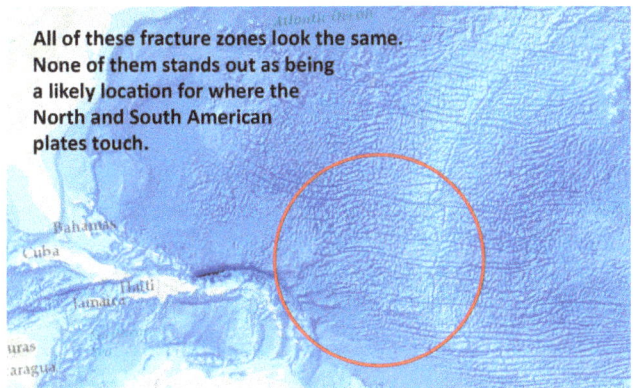

All of these fracture zones look the same. None of them stands out as being a likely location for where the North and South American plates touch.

The lack of convergent zones for the North American plate is a major problem for the theory. As Wikipedia's article on this plate tactfully puts it: "The motion of the plate cannot be driven by subduction, as no part of the North American Plate is being subducted, except for a small section comprising part of the Puerto Rico Trench; thus other mechanisms continue to be investigated."

Theorists have attempted to address some of the divergent/convergent issues by dividing up the plates into smaller plates. However, this has not been the panacea everyone expected, and many problems still remain. No convergent margins can be found for the Gorda, Juan de Fuca, and the Cocos plates, and the Manihiki and Magellan plates have neither spreading centers nor convergent margins.[6]

REASON #2: There is no adequate explanation for how subduction starts.

The theory of plate tectonics relies heavily on the idea of subduction. Diving plates are seen as the key to explaining what causes continents to drift.

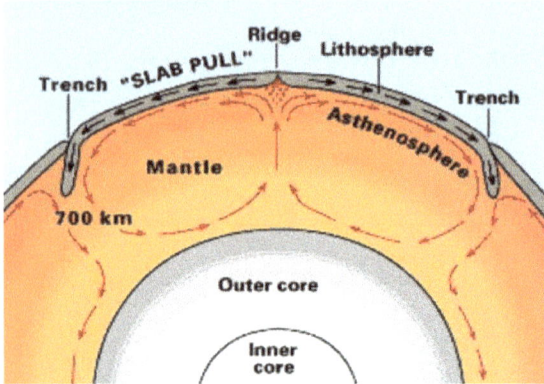

This graphic comes from the US Geological Survey.

As a "slab" of oceanic crust dives down into a trench zone, it is thought to pull the entire plate along with it. This has become known as "slab pull." Giving some extra help to slab pull we also have "ridge push." It is thought that at mid-ocean ridges, where the lithosphere is lifted up by the underlying asthenosphere, a plate might slide down the slope, being pulled by gravity. Thus, the word "push" might be slightly confusing, since the force being applied is a pull by gravity. The original definition of ridge push involved magma welling up at the ridge. It was thought that the magma was capable of pushing the plate. This idea has been largely abandoned, but the term "ridge push" was in use for so long that it continues to be used.

Though there are serious problems with slab pull and ridge push, the focus in this section is on the problem of how a plate first begins its dive. How does a massive piece of the earth's lithosphere push its way down through tens of miles of solid rock? Is this physically possible? Can using static analysis give us any clues?

Oceanic crust (the outer gray layer in the diagram above) has an average density of 3.0 g/cm³. The density of the underlying asthenosphere is estimated to be about 3.3 g/cm³. Less dense substances are found on top of more dense substances. Everyday examples of this are ice floating in a glass of water, or an air mattress floating in a swimming pool. Therefore, the structure of the crust and upper mantle are as we would expect, with the less dense rock on top.

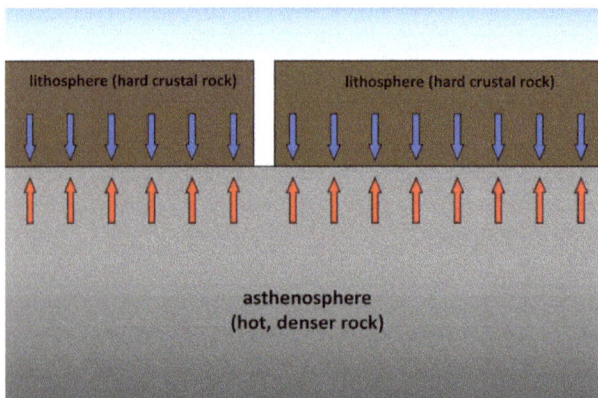

Here we see an engineer's diagram showing the forces that will be important in our analysis. The force of the lithosphere pressing down must equal the force of the asthenosphere pushing back up. If the forces suddenly became unequal, the rock will move up or down.

We can adapt the diagram to show one of the lithospheric plates beginning to dive down. The engineer's arrows show us the forces involved in this scenario. A

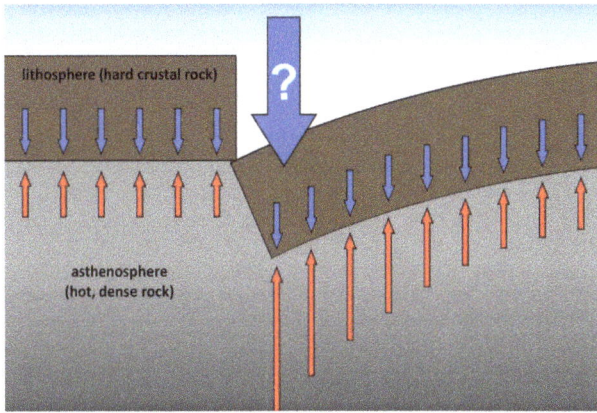

very large force from above must push the less dense rock down. The deeper the plate goes, the stronger the "push back" will be (indicated by longer red arrows). The force indicated by the large blue arrow would have to be unimaginably large. What could possibly provide this force? One hypothesis is that as the seafloor moves away from the ridge and works its way across the ocean over millions of years, it begins to thicken and to accumulate sediments. Perhaps by the time it reaches a subduction zone it will have become more dense than the underlying rock, and will be loaded with enough heavy sediments that it will be able to sink. Despite the fact that there isn't any supporting evidence for this idea[5], let's assume that this is plausible and has caused the plate in our diagram to dip down. But first, we need to look at another problem.

Granite is one of the most studied rocks in the world. We know not only its density and chemical composition, but also its compressive strength. We know how much pressure a piece of granite can take before it cracks. We also know that when granite is under pressure deep underground (over 5 miles [8 km] deep) its texture will become putty-like and it will slowly creep and flow if given the opportunity. This behavior of granite at great depths was observed in the world's deepest hole—the Kola "Superdeep" borehole in the Kola peninsula, east of Finland. The drillers found that they could not go deeper than about 12 km, because every time they brought the drill up for repairs, the bottom kilometer of the hole would close up. The pressurized granite flowed into the empty space as the drill was removed.

Now let's look at the scale of our diagram. These plates are 30-60 miles (50-90 km) thick. The arrows indicate that the granite at these depths will have a plastic consistency and will flow into any cracks that open up.

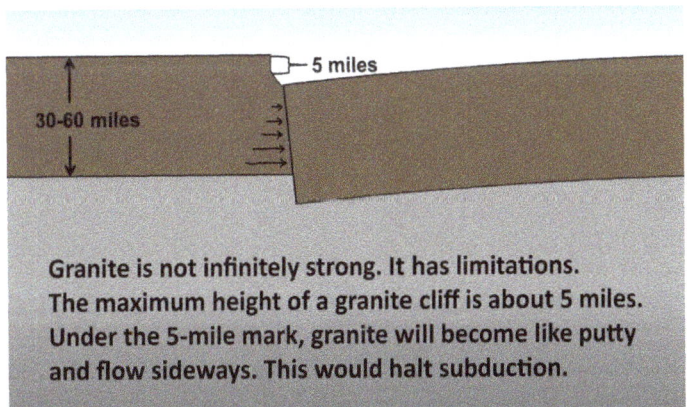

Granite is not infinitely strong. It has limitations. The maximum height of a granite cliff is about 5 miles. Under the 5-mile mark, granite will become like putty and flow sideways. This would halt subduction.

Let's be optimistic and assume that a plate has been able to overcome the squishy granite problem and has also somehow been able to collect enough sediments on its surface to push it down a bit. We'll ignore the fact that all seismic data suggests that the asthenosphere into which the plate is diving is more dense than the plate, and thus remove any problems with density. Would the plate actually be able to subduct?

This is how an engineer might choose to set up the problem,[7] using a method called static analysis. There are mathematical formulas that allow us to calculate all the forces involved. The green arrows represent the pushing force of the subducting plate. The red arrows represent the "push back" forces that the plate must overcome in order to subduct. If the sum of all the green forces is greater than the red ones, the plate will subduct. If the sum of the red forces is greater than the green, the plate will not subduct.

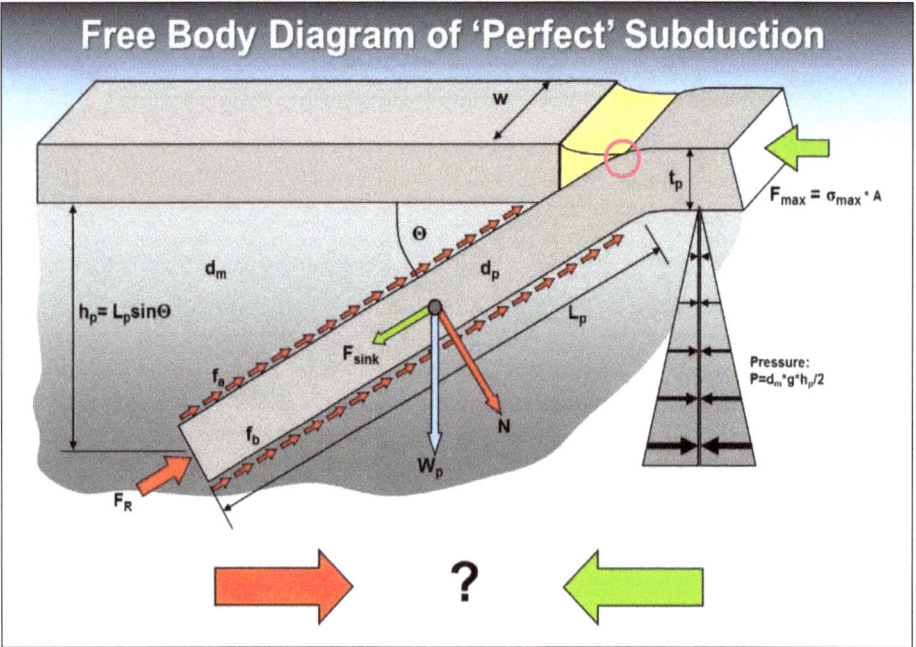

Free Body Diagram of 'Perfect' Subduction

Before we start calculating, let's make it easier for the plate to subduct. We will remove that red arrow pushing against the blunt front edge of the plate. This is like assuming that there is no rock at all in front of that edge. Also, let's assume that the plate won't crack at the point where it is bending down (marked by the pink circle).

We'll need to keep the estimates of the pushing and pulling forces to reasonable levels that won't exceed the strength of the rock. The pushing force (compression) can't crush the rock or turn it into putty. The downward pulling force (tension) can't cause the rock to tear apart. We will set these numbers to the maximum possible, using figures that were derived from experimental data.

You don't need to understand all the math in order to get the general idea of what is going on here. The tiny red arrows represent the frictional forces on the top and bottom of the plate. The plate width, "w" (the dimension that comes out toward you),

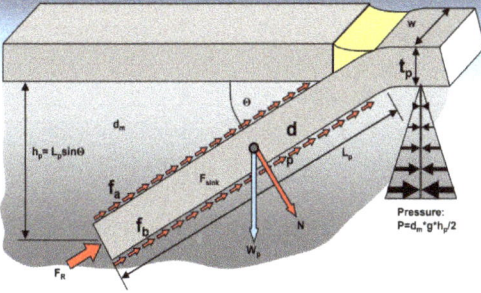

Gravity: $g = 980 cm/s2$
Max Compressive
Stress of Rock: $\sigma_{max} = 1.3 \times 10^9$ dyne/cm²
Mantle Density: $d_m = 3.2 gm/c\ m^3$
Plate Density: $d_p = 3.5 gm/cm^3$
Plate Length: $L_p = 35 \times 10^6 cm$
Plate Thickness: $t_p = 8 \times 10^6 cm$
Width of Subduction: $w = 1 cm$
Angle of Subduction: $\Theta = 30°$
Coeff. of Friction (rock-on-rock): $\mu = 0.6$

Add the Friction Forces $=$ f_{above} $+$ f_{below} $+$ ✗

$\mu*(P \quad * \quad Area) \quad + \quad \mu*(P \quad * \quad Area \quad + \quad N)$

$\mu*[(d_m*g*h_p/2)*(L_p*w)] \quad + \quad \mu*([(d_m*g*h_p/2)*(L_p*w)] \quad + \quad [d_p*g*(L_p*w*t_p)\cos\Theta])$

= 16.500 trillion Newtons
= 3.7 trillion lbs-force

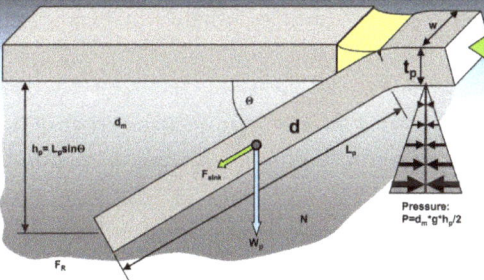

$F_{max} = \sigma_{max} * Area$

Gravity: $g = 980 cm/s2$
Max Compressive
Stress of Rock: $\sigma_{max} = 1.3 \times 10^9$ dyne/cm²
Mantle Density: $d_m = 3.2 gm/c\ m^3$
Plate Density: $d_p = 3.5 gm/cm^3$
Plate Length: $L_p = 35 \times 10^6 cm$
Plate Thickness: $t_p = 8 \times 10^6 cm$
Width of Subduction: $w = 1 cm$
Angle of Subduction: $\Theta = 30°$
Coeff. of Friction (rock-on-rock): $\mu = 0.6$

Add the Subductive Forces $=$ F_{max} $+$ F_{sink}

$\sigma_{max}*Area \quad + \quad W_p*\sin\Theta$

$\sigma_{max}*(t_p*w) \quad + \quad [(d_p-d_m)*g*(Lwt)]\sin\Theta$

= 515 billion Newtons
= 116 billion lb-force

is set to 1 cm not to reduce friction, but to make the math easier. Friction is not an area-dependent property, so we might as well keep the math as simple as possible.

Our final result is 116 billion (.116 trillion) pounds of force working *for* subduction, compared to 3.7 trillion pounds of force working *against* subduction. The anti-subduction force is over 30 times greater, even in an unrealistic, best-case scenario.

An objection at this point could be that this analysis doesn't take into account the fact that there could be magma under the plate, acting as a lubricant. Lubrication would turn this into a fluid mechanics problem, with a different set of definitions and equations. But where would the magma lubricant come from? It is true that when rock rubs against rock, the friction can cause melting. Once a plate is in motion, we can invoke magma as an additional pro-subduction force. However, we are looking at the question of how a plate *begins* to dive, and therefore, by definition, we don't have any motion yet. Until a plate is in motion and producing magma, we have to deal with the friction problem.

Plate tectonic advocate Agust Gudmundsson wrote an article for "Frontiers in Earth Science" in 2013[8] that summarized some of the challenges that the theory still faces. In the section dealing with initiation of subduction he gives the two leading ideas about initiation: 1) Plates age as they move away from divergent zones, and as they age they thicken. They also collect a heavy layer of sediments on their surface. 2) Convergence is caused by compressive force, often at a place where there is already a weakness in the crust, such as a fracture zone or continental margin.

If the first idea was true, he says that we should expect to find the subduction zones at roughly constant distances from nearby ocean ridges. What we actually find is that the distances vary widely. Also, when we go looking for examples of old lithosphere that has collected a massive load of sediments and is ready to subduct, we can't find any actual locations to use as prime examples. This idea has apparently fallen out of favor.

A major problem with the second idea is that there are numerous examples of fracture zones that are assumed to be under quite a bit of ridge-push stress and none of them has turned into a subduction zone. Researchers have located fracture zones with dips of 50-70°—ideal sites for subduction to start—and yet not one of these has any evidence of subduction. Also, the edges of the continental shelves in the Atlantic Ocean should be sites where the crust is prone to developing weak spots that could turn into subduction zones. What do we actually find? No subduction.

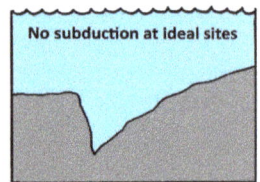
No subduction at ideal sites

Gudmundsson concludes that "while these and related models certainly indicate several possibilities as to subduction initiation, it is fair to conclude that we do not yet have a general quantitative theory—one with a clear explanatory power—as to subduction zone initiation." So many researchers have come to similar conclusions that the Wikipedia article on subduction admits that "the process by which subduction is initiated remains a matter of discussion and continuing study."

REASON #3: The Philippine plate has a noticeable lack of divergence.

The Philippine plate is wedged between the Pacific plate and the Eurasian plate. The long arrows show the direction of plate movement. The wavy black lines are the subduction zones and the tiny arrows along these lines show the direction of subduction, with the arrows pointing towards the stationary plate (under which subduction is happening).

The Philippine plate is almost entirely surrounded by convergent zones that contain the deepest trenches in the world. What is providing this plate with enough oceanic crust for massive subduction on all sides?

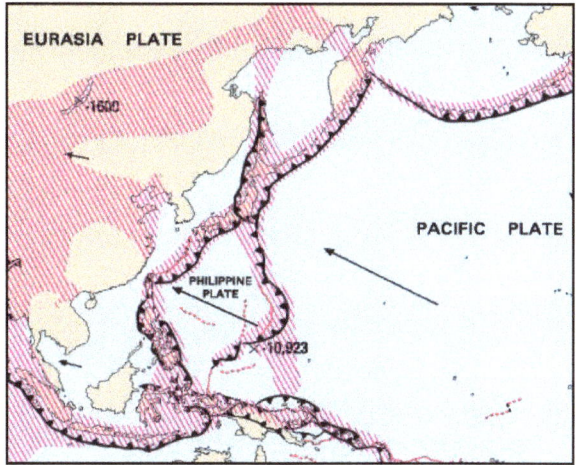

USGS map, public domain, found here: https://www.usgs.gov/media/images/pacific-plate-boundaries-and-relative-motion

This problem began when the theorists decided that trenches must *always* be convergent zones. By doing this, they "painted themselves into a corner" and ended up with an awkward situation in the western Pacific. In an attempt to make sense of this trench-saturated part of the world, they began to add tiny divergent lines wherever they could, and they subdivided the large plates into smaller ones. They added a divergent zone inside the Philippine plate (at letter A) along a ridge that contains many islands, even though this ridge looks nothing like the mid-ocean ridges. Adding this bit of divergence may have downgraded the problem from "theory-killer" to "major inconvenience," but it didn't actually solve the problem. Are we being asked to believe that the divergence at letter A is providing all the crust for the entire plate? What is happening at the northern tip of the plate where we have the subduction zones almost touching? How does this work?

File:Tectonic_plates_boundaries_detailed-en.svg created by fr:Utilisateur:Sting under CC-BY-SA, CC BY-SA 3.0, https://en.wikipedia.org/w/index.php?curid=20848040

At the bottom of the map, we see an attempt to make the theory work by creating smaller plates and more zones of all kinds. Critics of PT theory see this as an indication that the basic tenets of the theory need to be reassessed.

17

REASON #4:   The Pacific Ocean doesn't have enough distinct spreading centers, even though it covers a large section of the globe.

The Pacific plate has the same problem as the Philippine plate, except on a much more massive scale. The Pacific Ocean is, for the most part, surrounded by convergent zones. The only significant divergent zone is a very low and broad "rise" that defines its southeastern border. The East Pacific Rise (EPR) is the tag end of the mid-ocean ridge system. If you follow the rise to the west you will eventually end up at the Mid-Atlantic Ridge. The existence of the rise was known in the 1960s, but there weren't any detailed maps of it. In fact, there were very few detailed maps of any part of the Pacific.

In the 1960s, less than 0.0001% of the deep ocean had been explored. Even by the mid-1990s, only about 3-5% of the ocean basins had been explored in detail.[9] What was known about the Pacific in the early 1960s? Almost all of the Pacific islands had been discovered and mapped. The Aleutian Trench had been discovered and charted, and much was known about the floor of the Gulf of Alaska. The Ring of Fire and its association with earthquakes was well established. Hugo Benioff had published his work on the Tonga and Peru-Chile trenches and the seismic zones underneath them that dipped down at about a 45 degree angle. Many seamounts (extinct underwater volcanoes) had been discovered in the western Pacific, as well as many flat-topped seamounts that indicated a lower sea level at some time in the past. Some underwater mountain chains had been found. Magnetic anomalies (magnetic "striping") had recently been found along the coast of California. Also off the coast of Califor-

nia, SONAR had revealed non-seismic (not causing earthquakes) fault lines perpendicular to the coastline. The largest of these would later be named the Mendocino Fracture Zone. The East Pacific Rise had been found decades previously, but had been classified as a shoal. The Chile Rise was still uncharted, but it is marked on this map because it will help the theory.

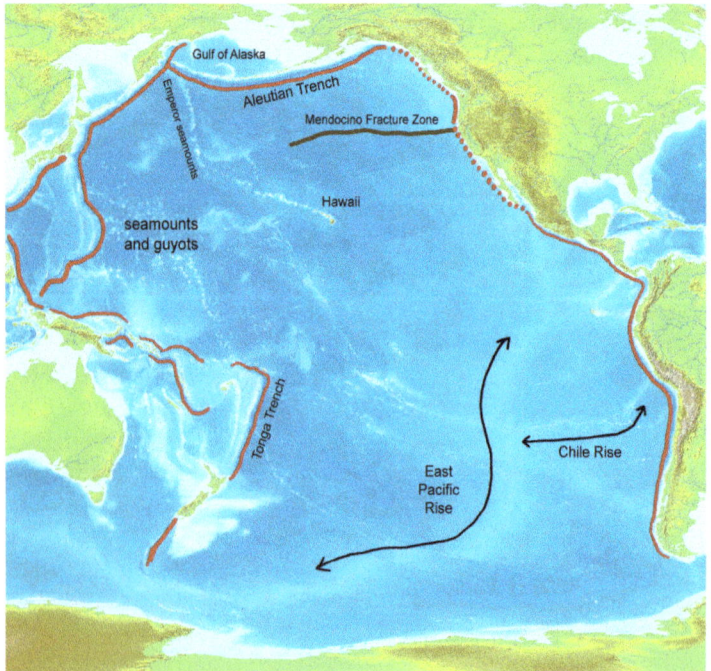

CC BY-SA 3.0, https://commons.wikimedia.org/w/index.php?curid=117566 (labels added)

The data available in the early 1960s already showed how different the Pacific was from the Atlantic. Looking at the Mid-Atlantic Ridge, it is easy to see why the theorists guessed that something was happening (or had happened in the past) that was causing (or had caused) the continents to be pushed apart. There might be a few enigmas here and there that would be hard to explain and would require some tweaks to the theory, but overall, the visual evidence was so compelling that the theory passed inspection "with flying colors" (an apropos nautical expression from the days when sailing ships from the Atlantic came into harbor with their flags flying, indicating they'd had a successful mission).

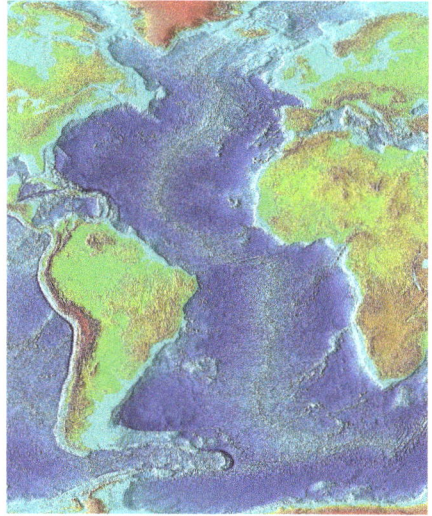

Why were the Atlantic and Pacific so different? Why was there almost nothing but divergence in the Atlantic and almost nothing but convergence in the Pacific? No one had a good answer. The early dissenters said that this was a red flag that something was fundamentally wrong with the theory, but the PT theorists rolled on ahead.

There was only one option: the East Pacific Rise must be the primary spreading center for the Pacific even though it was far from the center. The theorists trusted that as time went on, more spreading sites would be found, but until then, the EPR was "it."

If the EPR was spreading on both sides, that would mean that the eastern side would be going towards south America. Fortunately, the entire western coast of South America was one long trench, so getting rid of that crust wasn't a problem. Things got even better as earthquake data came in from that part of the ocean. They found a small line of minor quakes along what came to be known as the Chile Rise. Since earthquakes are used as primary evidence of a plate boundary, a line could be drawn there. More research found a few minor quakes happening along some small fracture lines west of Columbia, so another line could be drawn, finalizing what we now call the Nazca plate. This was seen to be a significant validation of the theory.

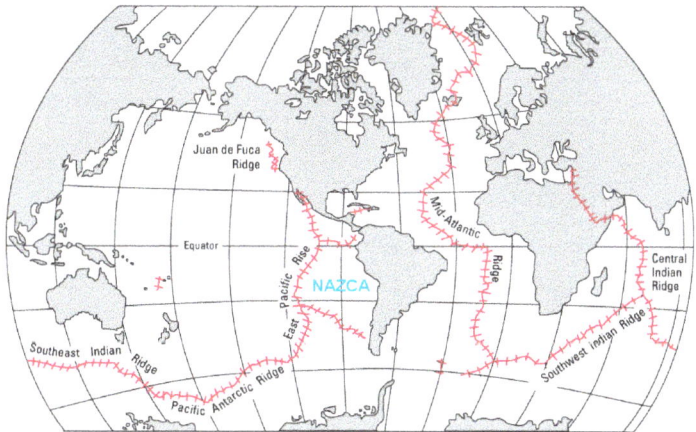

19

Even with such a tidy solution on the eastern side of the EPR, most of the Pacific Ocean lies to the west and north of the rise. (The southwestern part of the rise can be called the Pacific-Antarctic Ridge, but has historically been known as the western part of the EPR.) Is it physically possible for the EPR to push that much crust, even if it has some help from slab pull on the other side? The Pacific Ocean is thousands of miles across and has many islands and underwater mountains. Are we expected to believe that the relatively thin oceanic lithosphere wouldn't buckle or break under all the compressive (ridge push) stress? Shouldn't we see an underwater convergent mountain range forming parallel to the EPR?

There are also problems with the EPR itself. Research done since the 1960s has revealed that the EPR is very unlike the Mid-Atlantic Ridge. The MAR is a huge mountain range with tall peaks and many large perpendicular fracture zones. The MAR can rise 1,000 feet per mile in some areas. The EPR, on the other hand, has an average rise of only 10 feet per mile. If you were hiking up the EPR, it would seem almost flat. There are also very few perpendicular fracture lines. On Google Earth, the EPR is hard to locate even if you know where to look. It doesn't look like a place that is spewing out new seafloor. It also isn't high enough to give any gravitational help to "ridge push" forces. Along the MAR, gravity is invoked as an additional force that can help seafloor spreading by allowing the plates to roll away from the ridge. Gravity can't be a factor at the EPR.

Another solution that theorists have used to help solve the divergence problem is to look for more plate boundaries. More detailed maps of the seafloor have allowed convergent boundaries to be subdivided into small sections that alternate between convergent, divergent, and transform. A few small lines of divergence were added by these subdivisions. In this map, blue lines are subduction zones, red lines are spreading centers, and green/purple lines are places that fall into the transform category either because they are "strike slip" fault lines (sliding back and forth) or are ambiguous and require more data to be classified. This map of the area around Borneo and

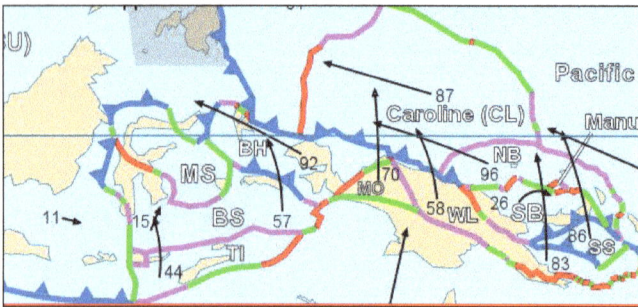

New Guinea shows how complicated these minor divisions have become. Indeed, they did find some additional mileage to classify as divergent, but these gains are tiny when compared to the overall problem.

The west coast of North America was only able to add a few hundred miles of divergence. As you can see from the map on the previous page there is a small area of divergence off the coast of Vancouver. This ridge (Explorer Ridge) is a geologically active site and looks much like the MAR. A small bit of success, but still minuscule in light of the overall problem. Most of the coast is classified as transform, with the plates sliding laterally. The San Andreas fault line is part of this transform boundary and is one of the most studied boundaries in the world. It is definitely not a speading center.

One of the original theorists, H. W. Menard, was aware that the lack of divergent zones in the Pacific would make the theory less viable. In 1964, he proposed that there had been an ancient spreading ridge in the middle of the Pacific that had subsided over millions of years and was now barely visible. He named the area The Darwin Rise, and hypothesized that millions of years ago the mantle began to bulge in this area, creating massive volcanism that was unique in the geological record. According to Menard, these volcanoes created such massive outpourings of lava that the entire seafloor was covered (so all features now in that area are the result of this event), then after the area cooled, subsidence occurred because the mantle had transferred so much of its volume to the surface.[10]

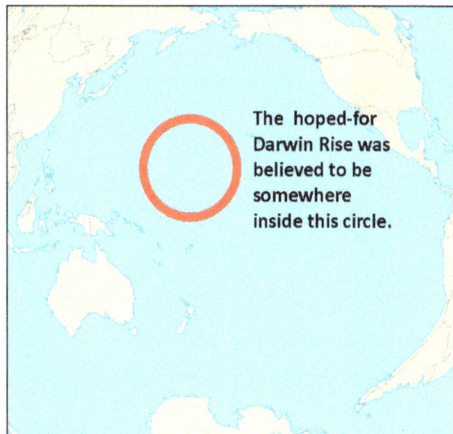

The hoped-for Darwin Rise was believed to be somewhere inside this circle.

By Tentotwo - Own work, CC BY-SA 3.0, https://commons. wikimedia.org/w/index.php?curid=18241876

Subsequent exploration of this area found that the seafloor was at a normal depth and that the guyots in the center were shorter than those in the periphery. The shorter guyots could not be explained by Menard's hypothesis. In 1984, Menard published again, restating his thesis (calling it the "Darwin Rise Reprise") but restricting its geographical area and adding a new idea—that there had been a mantle plume (an upwelling of deep magma) in this area, similar to the one conjectured to be under the East Pacific Rise area.

The existence of the Darwin Rise was largely unchallenged until 1989, when bathymetry for this area was carefully reanalyzed by Christian Smoot and Robert King.[11] Detailed cross-sectional profiles of the area were examined and found to be dissimilar to all the established ridge and rise areas of the oceans. No evidence of a "fossil" ridge could be found. The heights of the guyots did not form a pattern that indicated this area was, or ever had been, a mid-ocean ridge. The magnetic patterns on the seafloor in this area were not bilaterally symmetric, as is the case with all ridge areas, but instead formed a triangle. (This led to a new hypotheses that perhaps the Pacific had started here, as a small triangular sea.) The researchers concluded that Menard had based his hypothesis on early data that had picked up some points of higher elevation, but did not give a clear picture of what the area actually looked like.

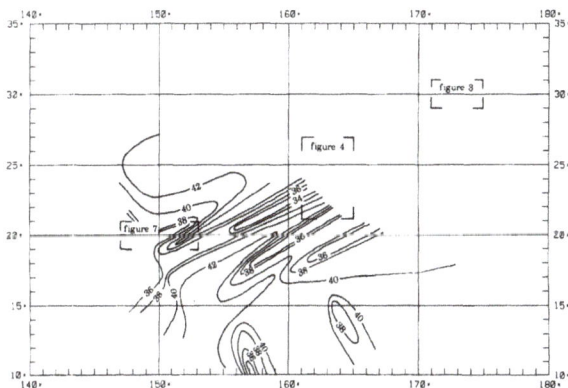

Smoot and King's contour map of the proposed "Darwin Rise" area, showing not a rise but a continuation of a fracture zone.

21

REASON #5:   If subduction has happened, we should find a huge volume of sediments that have been scraped off as the plate subducted. Instead, we find shallow, relatively undisturbed sediments.

Most geology textbooks state that as a plate subducts, it will drag a substantial amount of water and sediment down with it. Sediments will also pile up, or "accrete" along the edge of a continent, forming "accretionary wedges." The problem of whether it is possible for sediments and water to go down with a plate will be set aside for the moment. In this section, we will calculate how much sediment lies on the ocean floor, and how large the pile would be if a fair portion of it ended up at the edge of a continent. We will also contemplate the fate of underwater mountains (seamounts) as they ride along on top of the moving plate.

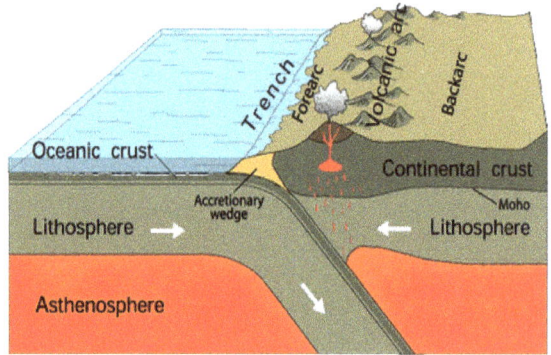

The floor of the Pacific covers an area of about 155 million square kilometers. (This figure is from a NOAA website. Estimates can vary quite a bit.) The depth of its sediments range from 300 to 600 meters (as listed on britannica.com). Assuming that seafloor spreading has been going on for hundreds of millions of years, how much sediment has been transported into the convergent zones on the western side of the Pacific (shown in red on the map below)? Let's assume a best-case scenario and keep our estimates very low, and let's keep the math simple. We'll estimate the area of the Pacific to be only 100 million square kilometers, and we'll take the low end of the sediment load, 300 meters. Every square kilometer of ocean floor would have .3 cubic kilometers of sediment, giving the entire ocean a total sediment volume of 30 million cubic kilometers. Let's say we've overestimated, and bring that down to only 10 million cubic kilometers. Now we need to distribute this sediment load along the 13,000 kilo-

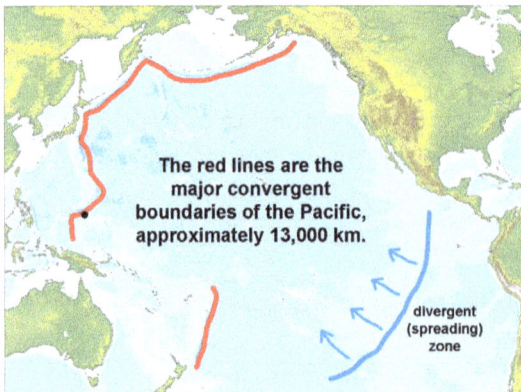

The red lines are the major convergent boundaries of the Pacific, approximately 13,000 km.

divergent (spreading) zone

meters of convergent boundaries (mostly trench zones) located in the western Pacific. Dividing 10 million by 13,000, we find that every kilometer of the convergent boundary has had to deal with about 770 cubic kilometers of sediments. It's hard to imagine the size of this pile. What would it look like if we dumped it into the world's largest ocean trench? (The deepest trench is the Mariana Trench, located at the black dot.)

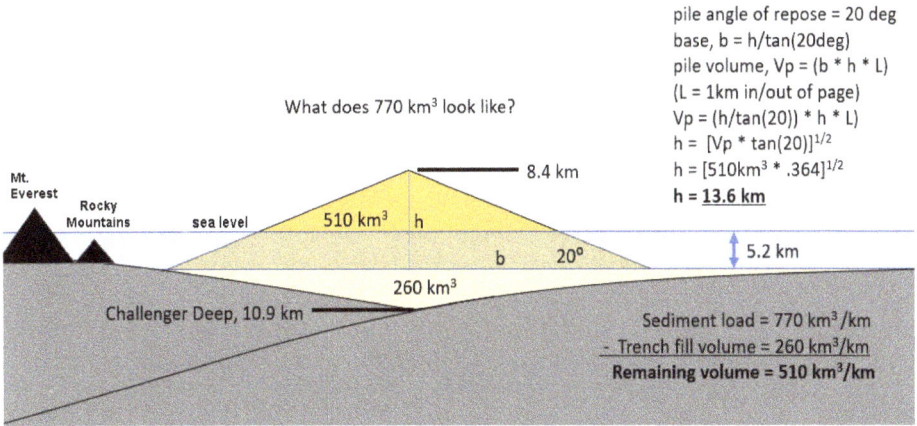

What does 770 km³ look like?

pile angle of repose = 20 deg
base, b = h/tan(20deg)
pile volume, Vp = (b * h * L)
(L = 1km in/out of page)
Vp = (h/tan(20)) * h * L)
h = [Vp * tan(20)]$^{1/2}$
h = [510km³ * .364]$^{1/2}$
h = 13.6 km

Mt. Everest
Rocky Mountains   sea level   510 km³   h   8.4 km
b   20°   5.2 km
260 km³
Challenger Deep, 10.9 km

Sediment load = 770 km³/km
- Trench fill volume = 260 km³/km
Remaining volume = 510 km³/km

The Challenger Deep is the deepest part of the deepest trench in the world, the Mariana Trench. Mt. Everest and the tallest peak of the Rocky Mountains are super-imposed onto this diagram to help us understand the scale of this pile. Even if half the sediments somehow managed to go down with the plate, we'd still have a pile much larger than the world's biggest mountains. (And don't forget how conservative we made our initial estimates.)

What do real accretionary wedges look like? One of the best examples is the convergent zone just south of the Aleutian Islands. This is considered a classic, text-book example of an accretionary wedge. (The "terrace" is the wedge.)

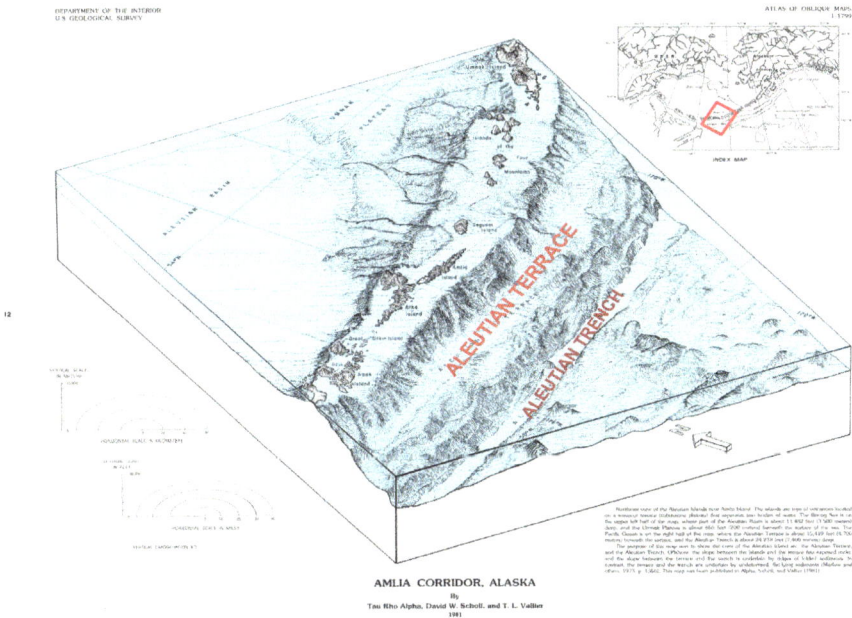

AMLIA CORRIDOR, ALASKA
By
Tau Rho Alpha, David W. Scholl, and T. L. Vallier
1981

Illustration taken from the <u>Atlas of Oblique Maps</u>, published by the U.S. Geological Survey.

23

The terrace. which is assumed to be the accretionary wedge, is far below sea level and is about the size you'd expect a wedge to be if subduction had started relatively recently. It is obviously not the accumulation of 770 km³ of sediments.

An unexpected addition to the sediment problem came when research on the sediments in the Aleutian trench revealed that the vast majority of them are not from the Pacific seafloor but from glacial erosion in Alaska. Chemical analysis of the sediments clearly showed that they are the result of erosion of continental rocks. This eroded material had washed down into the Gulf of Alaska, eventually ending up in the trench.[12] Continental sediments would have to be added to the 770 km³.

A partial solution to the sediment problem has been suggested: deep water currents that run along the trench "rework and redistribute material supplied from the accretionary prism and the oceanic plate."[13] If this is true, it just moves the sediment problem to another location. Kick the can down the road and it's gone, right?

In 2021, a group of European scientists published a lengthy paper in *Geosphere* (published online by the Geological Society of America) about accretionary wedges.[14] They proposed that a seamount is able to descend along with the plate, acting like a bulldozer, pushing a load of sediment. Not only does this help to get rid of some of the vast amounts of sediment, but it also provides a possible mechanism to explain uplifted areas along coastlines and earthquakes and volcanic activity characteristic of coastlines around the Pacific rim. This illustration shows their idea.

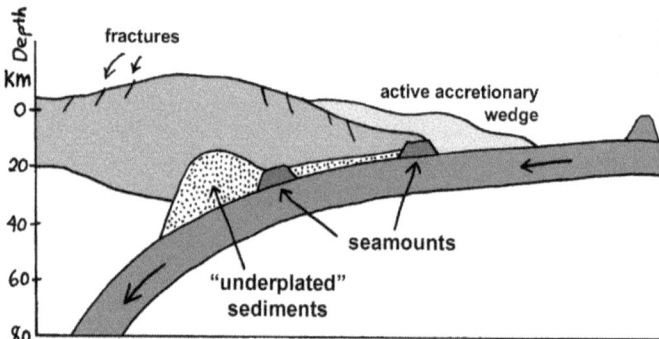

Is it reasonable to think that a cone-shaped object would be able to act like a bulldozer? (Imagine pushing an upside down ice cream cone through wet sand.) Wouldn't the sediments slide around the cones? Also, wouldn't the lumpy seamounts act like brakes and slow down the subducting plate? And would a pile of sediments really be able to lift a 10-20 km thick continental plate made of granite? The path of least resistance here would be to the sides. If the subducting plate is already deforming the warm rock through which it is moving, why should we think it would have any effect on the colder, more solid rock above? Wouldn't the warm rock move laterally?

All the action shown in the diagram is thought to be happening at depths of 10 to 50 km, which is deeper than we are able to drill for core samples. Thus, the only data we have comes from seismic studies. Sound waves are sent deep underground and the echo is recorded. When the echo results are processed by computers into a visual image, it looks something like the following diagram.

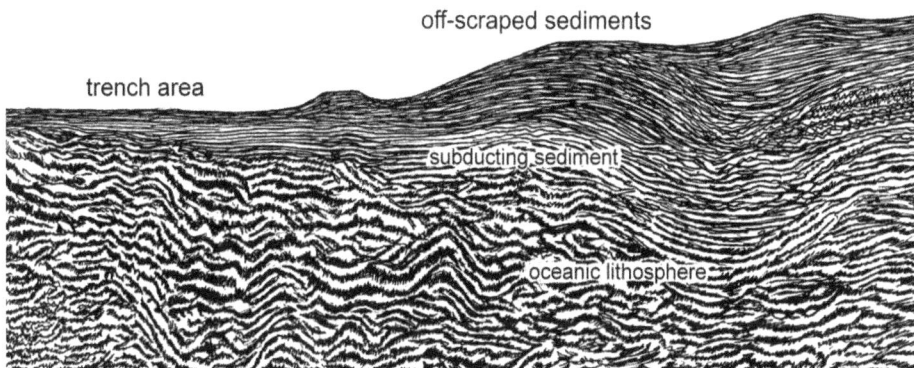

This illustration is an artist's rendering of actual seismic data (U.S. Geological Survey, McCarthy and Scholl, 1985) taken along a cross section of the western end of the Aleutian trench. The labels are those used by the authors of the paper. Being PT advocates, the authors had an interpretation already in their minds, and it was only a matter of figuring out where to paste in the labels. However, if you approached this image with no preconceived notion of what was going on, would you come to the conclusion that sediments were being subducted down into the mantle?

The sediment problem did not escape the notice of the early dissenters. In 1970, there was already enough information about the size and shape of trench zones that V. V. Beloussov[15] could write: "Given the rate of motion of 3 cm per year over 100 million years, and an average sediment thickness of 200 meters, the thickness of sediments piled up in a 50-km wide trench should reach 10 km high." About the nature of the sediments found in trenches he says, "Nowhere do we find those deformations in trench sediments which should, under such circumstances, arise. On the contrary, sediments occur undisturbed on the floor of trenches."

If subduction didn't happen, what did? Perhaps a scenario like this is more compatible with the seismic data shown above:

At some point in the past, the oceanic lithosphere was subjected to lateral compression and got a bit scrunched. Then, over thousands of years, sediments fell out of the water and blanketed the ocean floor, creating relatively flat, horizontal layers. Earthquakes in recent millennia have probably caused some additional deformation of both the lithosphere and the sediment layers.

25

REASON #6:   Seamounts (extinct underwater volcanoes) are too abundant and are often found in locations that don't align with PT theory.

Seamounts are defined as under-water mountains (assumed to be extinct volcanoes) that are 1,000 to 4,000 meters tall. Their base can be as wide as 100 km. Underwater mountains that are less than 1,000 meters tall are called "sea knolls."

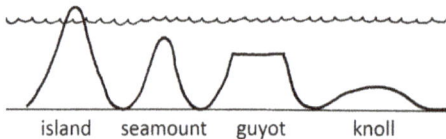

island   seamount   guyot   knoll

If a seamount is tall enough that it rises above the surface of the ocean, we call it an island. Most seamounts have pointed tops, but some have flat tops and are called "guyots" *(gee-ohs)*. Guyots show us that sea level used to be far lower than it is today. The guyots rose above the surface when they first formed, and were planed off by waves.

There are about 15,000 seamounts around the world. The red dots on these maps are seamounts. Green dots are guyots. Orange indicates continental shelves.

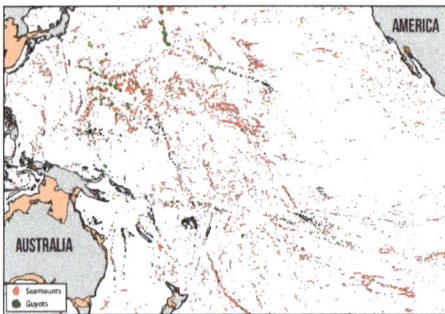

"Distribution of seamounts and guyots in the North Pacific"
by PeterTHarris, 2015, found in the Wikipedia article "Seamounts."

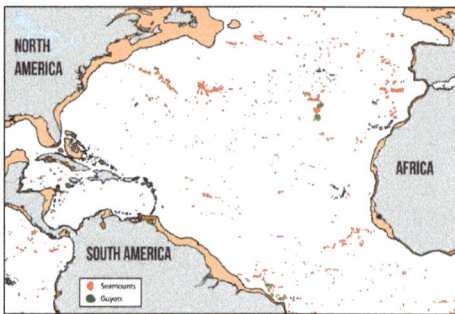

"Distribution of seamounts and guyots in the N. Atlantic"
by PeterTHarris, 2015, found in the Wikipedia article "Seamounts."

As you can see, the Pacific Ocean is littered with seamounts. Sometimes they occur along lines, but most often they are scattered in what looks like a random pattern. How did all theses seamounts form?

Geology textbooks[16, 17] tell us that volcanoes form for three reasons:
1) Subducting plates melt and produce lava that rises to form volcanoes on the landward side of a trench.
2) Spreading at divergent zones (mid-ocean ridges and continental rift zones) allows magma to come to the surface.
3) Plumes of magma ("mantle plumes") come up from the outer core, creating "hotspots" that release magma onto the ocean floor. Seamount chains are believed to be the result of a plate moving across hotspots.

Which of these reasons can explain the thousands of seamounts in the Pacific? Very few of them are close to subduction zones, and those that are close are often on the "wrong side" (the seaward side) of a trench, so the first explanation falls short. Since there are very few spreading centers in the Pacific, the second explanation can't be used in most cases. (The attempt to find remains of an ancient spreading center, The Darwin Rise, failed.) That leaves only the mantle plume explanation. The existence of hot plumes rising through the mantle is highly speculative, but even if magma could rise through the mantle and make it to the surface, is it reasonable to think that this has happened tens of thousands of times? Why would one side of the mantle (the Pacific side) produce many more plumes than the other side? Is the mantle unbalanced in some way?

The mantle plume explanation for "anomalous volcanism" (i.e. volcanoes far from any divergent or convergent zones) was originally proposed in 1971 by W. Jason Morgan. He thought there could be as many as 20 plumes in the mantle. Since then, the number of proposed plumes has grown to over 5,000. Leading researchers now say this number is far too high, but even if it is accurate, it still leaves thousands of seamounts unexplained. (The mantle plume controversy (including hotspots) is addressed in Reason #17.)

Here is a map that is very similar to several seamount maps you can find by doing an online image search. The size of the red dots exaggerates the actual size of the seamounts, but the map conveys the truth that seamounts are far too abundant to have been created only by the methods suggested by PT theory.

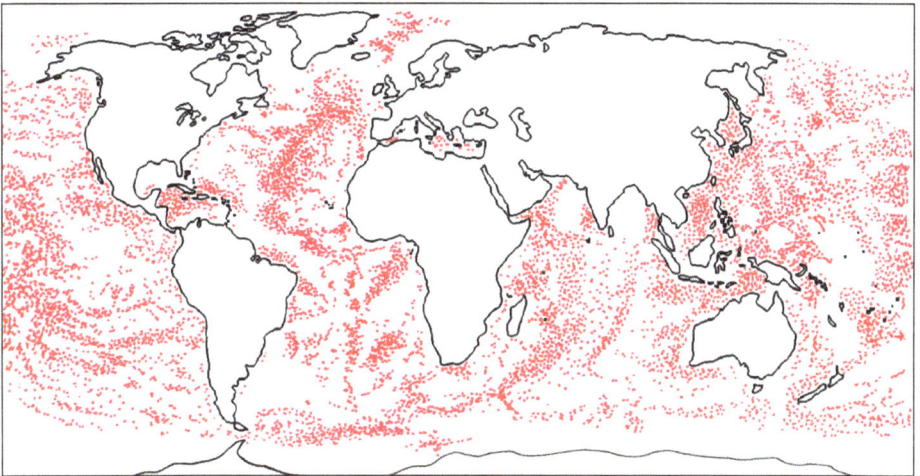

REASON #7: Seismic tomography has not shown unambiguous images of subducted plates.

Seismic tomography is a technique for imaging the inside of the earth using seismic waves produced by either natural earthquakes or manmade explosions. Sensors are placed at many locations and they each record the seismic waves as they come in. By comparing the data from all the locations, the direction, speed, and point of origin of the seismic waves can be determined. Several types of waves are generated by quakes and explosions. P-waves, which can travel through both solids and liquids and are the first waves picked up by sensors, are the waves used in seismic tomography.

P-waves (pressure/primary waves) travel faster through more dense materials. If you've ever put your ear on a table and had someone tap the other end, you've witnessed how efficiently a dense object transmits waves. Imagine putting your ear to a mattress, which is much less dense than a table. A tap on the other side of the mattress might not even reach your ear at all. Different types of rock have different densities, and a seismic sensor can pick up these differences. Lab experiments have recorded the behavior of seismic waves through various types of rock, so wave velocity can tell us a lot about rocks deep underground. However, other factors, besides density, can also affect how fast waves travel. Temperature is also a key factor, with waves traveling faster through cold rock than warm rock. The presence of water in or around the rock can also have an effect. (Reason #19 discusses the reliability of seismic data.)

How can we know whether faster P-waves mean colder rock or more dense rock? The truth is that we can't know, but that doesn't stop PT advocates from confidently asserting that you can see subducting plates on images made using P-wave tomography.

Here we see a well-known P-wave tomography image. Because it was released into the public domain, it appears in many online articles about plate tectonics. Data about an earthquake was gathered by many seismic sensors then compiled by a computer and put into a form that makes wave velocities very easy to see. The key at the bottom shows that the colors represent how much faster, or slower, than average the P-waves traveled. Blue is 3-6% faster.

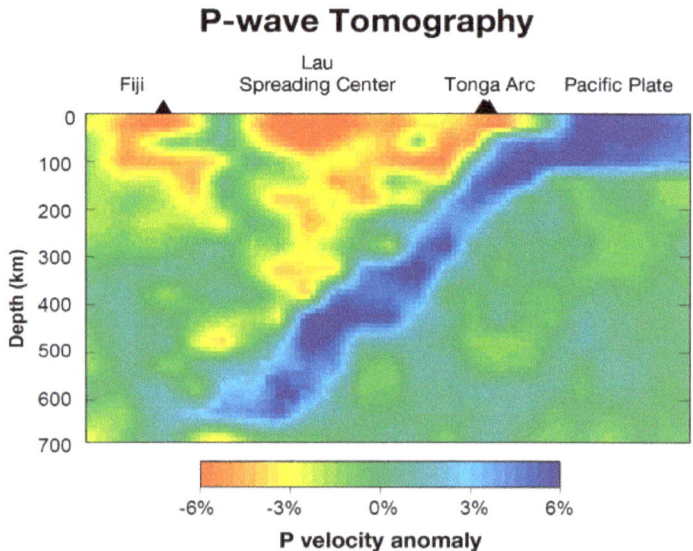

P-wave Tomography

The earthquake used to make the tomographic image occurred under the Tonga Arc, shown as a red line on this map. The image is a cross sectional view, shown by the black line. Imagine slicing down into the earth along that black line.

The diagonal blue shape in the tomography diagram is an area of rock where the P-wave velocity is 3-6% faster than the surrounding rock. (To be clear, P-wave tomography does not show anything moving. This is a still image, like a photograph.) PT advocates argue that this blue region is obviously a cold plate subducting into the mantle. But is there another, equally valid explanation? Could the faster P-waves indicate an area of higher density instead of colder temperature?

P-wave velocities through solids vary inversely as the square root of the absolute temperature.[18] Therefore, P-waves should travel about 400% faster through cold subducting plates, not the measured 6%.

If the blue area is a cold plate, we can estimate that its temperature as it begins to subduct is the same as the seafloor, about 3° C (275° K). At a depth of 700 km, the plate would be surrounded by the warm rock of the mantle. The mantle at that depth is estimated to be as hot as 1700° C (2000° K). It's a bit strange that we don't see much change in temperature as the plate dives to 700 km. Assuming that the plate is moving only a few centimeter a year, the tip of the plate has been down there for a very long time. You'd think that after (the assumed) million years of being in a 1700° C environment, there would have been some heat exchange. Also, notice the change in the shades of blue. Darker blue is interpreted to mean "colder." Why does the subducting plate start to get less cold at about 200 km depth, then go back to its initial cold temperature again at 400-500 km depth? How could it get colder if the surrounding rock gets hotter as you go down?

There is another equally valid interpretation of this image. Blue simply means that the P-waves go a little faster. This could be an area that is more dense than the surrounding rock. It is obvious that the Pacific plate is very fractured. At some point in the past, a large fracture occurred along this blue area. The faces of the rock along the fracture moved in opposite directions, causing a massive amount of friction (and likely a huge earthquake). Even a small movement of rock at this great depth can cause immediate and catastrophic melting. This slippage melted the surrounding rock, possibly causing even more slippage and more melting. Eventually, the melted rock cooled, but as it did, it became more dense. (Think of what happens to ice cream when it melts then refreezes. The refrozen ice cream becomes more dense, and is therefore not as pleasant to eat.) The blue area could be rock that melted then cooled, becoming more dense in the process. The 6% higher seismic velocity is exactly what you'd expect to find if the blue area was more dense due to melting and cooling.

REASON #8: Benioff zones do not prove the existence of diving plates. In fact, the more data that comes in, the more enigmatic they become.

In 1949, American scientist Hugo Benioff discovered how to use seismic data to determine exactly where the epicenter of an earthquake occurred, and how to plot the epicenters over time to show fault planes. Kiyoo Wadati of Japan made a similar discovery in the late 1920s, so these zones are often referred to as Wadati-Benioff zones.

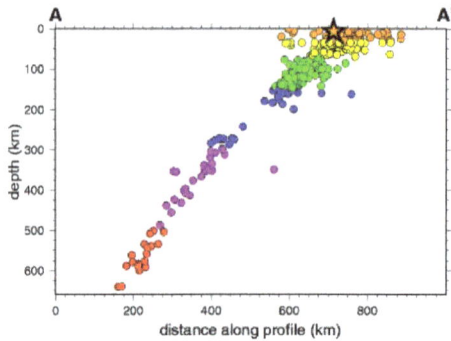

The dots in this graph are the epicenters of earthquakes under the Kuril Trench area north of Japan. The earth-quakes get deeper farther away from the trench. (The trench is marked by a star.) Does this graph "show" a subduct-ing plate? No, it shows the epicenters of earthquakes, presumably along a fault plane that slips frequently. Interpreting this as a subducting slab is an inference, not at observation.

PT advocates saw Benioff's work as proof of subduction. To them, that line was definitely a subducting slab. For several decades, this interpretation seemed to work. Then, more data started to come in. The first big surprise was the discovery of Double Benioff Zones (DBZs). In some Benioff zones, two lines of epicenters could be detected. Further research revealed another surprise: DBZs are not rare, but are a common, world-wide phenomenon.[19] PT advocates have proposed several explanations for the DBZs, such as sagging of the plate, the unbending of the slab, or the separation of different minerals over time, What they never consid-er is that these lines could simply show two fracture planes, not two plates.

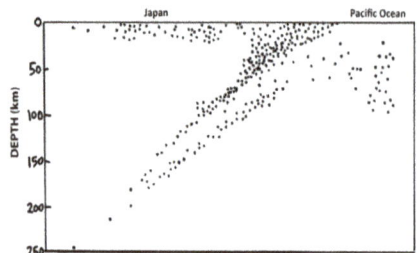

These lines of dots are interpreted as an upper and a lower plate (instead of simply two fracture planes).

The next surprising find was that some Benioff zones barely dip down at all. For example, the Benioff zone along the west coast of South America (assumed to be the sub-ducting Nazca plate) was discovered to be close to horizontal at a depth of about 100 km. They double-checked the data by setting up sensors over that area instead of relying on data from worldwide earthquake sensors. The new data showed that these "flat slabs" extend very far under the continent, even under northern Argentina.[20]

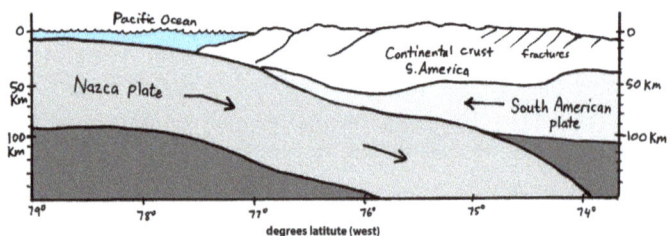

PT theory now had to find an explanation for these flat slabs. The first idea was that the subducting oceanic crust was made of rock that was not very dense, and therefore relatively buoyant—too buoyant to sink into the mantle the way other subducting slabs did. Additionally, perhaps "bathymetric highs" such as ocean ridges, plateaus and seamounts were also preventing the plate from sinking. However, other flat slabs were found in places that did not have bathymetric highs[21] and normally subducting slabs were found in places that did have bathymetric highs. Additional research using computer modeling made this solution look even more doubtful. [22]

The second idea involved both the shape of the over-riding plate and two new mechanisms they invented for this occasion: trench suction and trench retreat. If the over-riding plate has a large "keel" (a large bump on the bottom) perhaps this can impinge on the flow of the mantle near the trench and cause flattening. Trench suction could have several causes, they said, but would definitely play a role in the overall depth of subduction. Trench retreat is when a trench moves in a direction that is opposite to the direction of convergence and is a very convenient mechanism to invoke when the data doesn't seem to be lining up with your hypothesis. But not to worry, if you make a computer model and enter all the right initial conditions, you can get the program to indicate that trench retreat is possible, though only at the ends of a convergent zone.[23] (One can't help but think of that Procrustean bed analogy made by one of the early dissenters. PT theory can be pulled, stretched, or chopped to fit any data set.)

There is also much speculation about what might be happening to the minerals in the flat slabs as they are subducting. The rock of the oceanic lithosphere might undergo metamorphic changes and turn into a more dense version of itself, called eclogite. Perhaps flat slabs are places where this process was delayed due to the lack of water-retaining cracks along the bend at the trench area. This is hard to prove because these chemical changes would be happening at depths far below our ability to drill for core samples. We've barely scratched the top of the ocean floor in just a few places. We don't know for sure what type of rock lies under oceanic basalt.

Benioff zones weren't done surprising scientists. Studies of the Aegean[24] and the Tonga trench have shown that Benioff zones can be made of separate sections that dip at different angles. These sections can be of different lengths and there can be gaps between them. Are these gaps places where there isn't any slab, or are they places where the slab just doesn't produce any earthquakes? How does "slab pull" work in these areas? Is each section pulling at the same rate but in a different direction? Why don't these cracks propagate up to the edge of the seafloor where we can see them? And if ALL that seafloor is going down into the curved trench, shouldn't we see the excess rock as thick, folded slabs? Where is all the subducted rock going? These geometry issues will be explored in Reason #9.

This paper model approximates the computer generated images of the Aegean Benioff zones. The sections dip at different angles.

REASON #9: The arc-and-cusp pattern around the Ring of Fire and along trench zones creates an impossible geometry for plate subduction.

This problem won't require quoting scholarly research articles. It's based on common sense knowledge we all have about the way materials behave.

In textbooks, we are always shown a two-dimensional diagram of a perfect, "straight-line" subduction scenario.

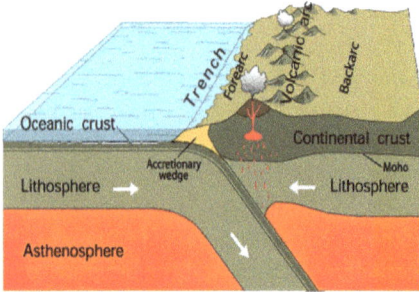

Here is the Wikipedia diagram again. Most textbook diagrams are similar.

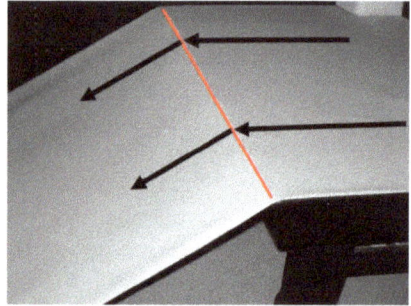

This demonstration simulates a straight-line subduction scenario, as in diagram.

The arrows on the demonstration model above show that the force vectors are straight and do not pose any problems as the plate bends down at the red line. As the white fabric goes over the bend, it folds down nicely and no wrinkles form. However, convergent zones (trench zones) are usually curved, not straight.

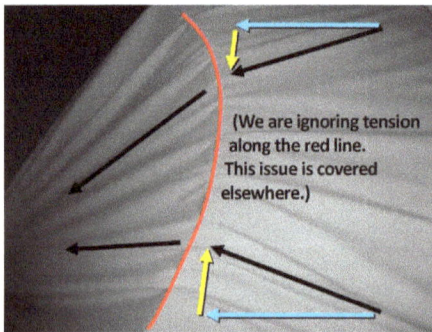

(We are ignoring tension along the red line. This issue is covered elsewhere.)

This demonstration model (left) represents a typical curved trench zone. As the white fabric is pulled over the red line, many wrinkles form. All flat surfaces will do this when they are pulled over the edge of a curve. The blue and yellow arrows show the force vectors involved. The yellow arrows show the lateral force that is squishing the material together as it subducts. The fabric (or rock) will necessarily have to fold and wrinkle.

When we look at the force vectors where two curves meet, we see another problem. The forces create a tensile region where material is being pulled in opposite directions. We should expect to find a tear, a split, a place where the rock has greatly thinned out, or a bulge of magma that has filled up the tear. Ocean floor exploration has not revealed any of these.

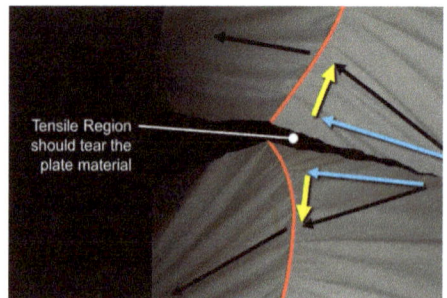

Tensile Region should tear the plate material

Along the Pacific rim we find a series of curves forming what is known as an "arc and cusp" pattern. The cusps are the points where the curves meet. What type of topography should we expect to see along these arcs, if plates really are subducting? Shouldn't there be geological wrinkles and tears? There are only a few features that might qualify as evidence of a tension tear at a cusp; they are marked with red arrows. However, there can be other equally plausible explanations for how these features formed. Essentially, as we look at the topography of the Pacific floor, we have reason to ask, "Where are the wrinkles?"

Red arrows show lines that might be used by advocates to prove that we do see some wrinkles.

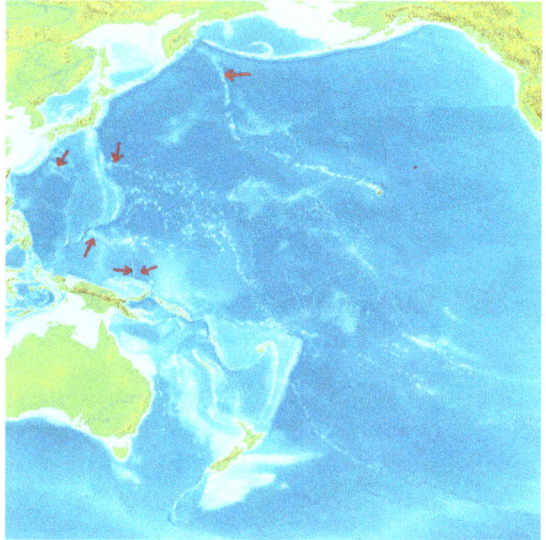

However, shouldn't we see "wrinkles" like these? (white lines)

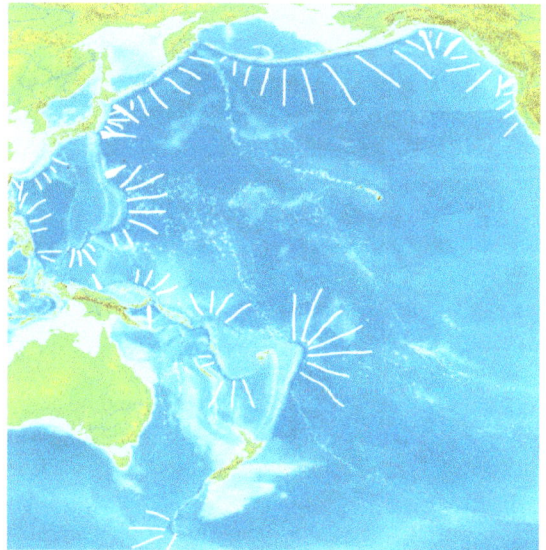

REASON #10: If the floor of the entire Atlantic Ocean was created by the Mid-Atlantic Ridge, why doesn't it have the same (or similar) texture as the ridge? Shouldn't we see a series of parallel ridges across the seafloor?

Vast areas of the Atlantic Ocean (the areas that are dark blue on this map) are known as "abyssal plains." They are deep and dark and very flat. There are isolated sea-mounts or clusters of seamounts in some places, but below the seamounts are some of the flattest places on earth. PT advocates believe that this ultra-flat ocean floor was "born" at the mid-ocean ridge millions of years ago. The mechanism by which the ridge texture is removed is never explained.

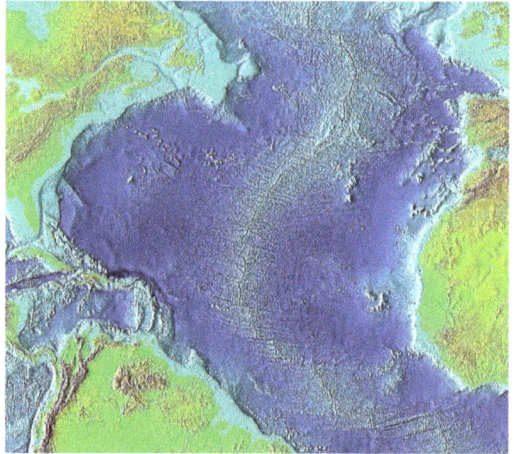

PT theory says that as magma wells up at the Mid-Atlantic Ridge, it displaces the rock at the ridge and pushes it away on both sides. Since lava cools quickly in water, it would not take long for those hot areas to cool off and become solid. In fact, even right at the middle of the ridge we don't find soft rock. Dredge hauls along the center of the ridge have brought up a surprising variety of rocks, including some chunks of continental rock. We don't find hot, soft rocks—ridge rock is cold and hard. If cold, hard, ridge rock has been traveling away from the ridge for millions of years, shouldn't the final result be a series of parallel "former ridges" all the way across the ocean on both sides? Eroded slightly, perhaps, but still visible because they would have started out several kilometers high (the height of the current ridge).

Is erosion capable of "erasing" all the former ridges? Would underwater currents be able to completely wear away rock mountains several kilometers in height? We don't see this type of erosion happening right now, so trusting that it somehow happened in the past is wishful thinking.

Perhaps we don't see these former-ridges because they started sinking into the mantle soon after they began to move away from the ridge. If so, why didn't they form a subduction zone right there, close to the ridge? Why did they wait so long to subduct? This problem was first pointed out by V. Beloussov in 1970.[25]

Are the former ridges completely covered by sediments by the time they reach the abyssal plains regions? If so, we should be able to detect them using seismic technology, as basalt is much more dense than the sedimentary layers and the ooze that blankets the ocean floor. So far, nothing looking like "former ridges" has been detected.

Judging by the cartoon diagrams of seafloor spreading that we see in textbooks, PT advocates seem to imagine the ridge as having the texture of hot asphalt or wet sand that slowly sags and oozes over time, allowing the ridges to gradually flatten out and disappear. It's amazing that this unrealistic assumption has gone unchecked for decades.

REASON #11: The uneven topography of the lithosphere/asthenosphere interface makes it impossible for continental plates to move.

Mountains have "roots." Seismic studies have shown that mountains have as much, or more, crustal mass under the surface as they do above. The layers of rock are said to be in *isotacy*. The crust and lithosphere are "floating" in the asthenosphere at the levels predicted by their relative densities. One study found that the crustal roots under the Himalayan mountain range in southern Tibet could go as deep as 80-90 km. (Average crustal depth in non-mountainous areas is about 30 km.) The roots of the Alps have been estimated at 65 km deep, while the mountain range itself rises only 4 km above sea level. Even some parts of the world that are not mountainous have deeps roots, such as Finland, which is sitting on crust as deep as that under the Alps.[26]

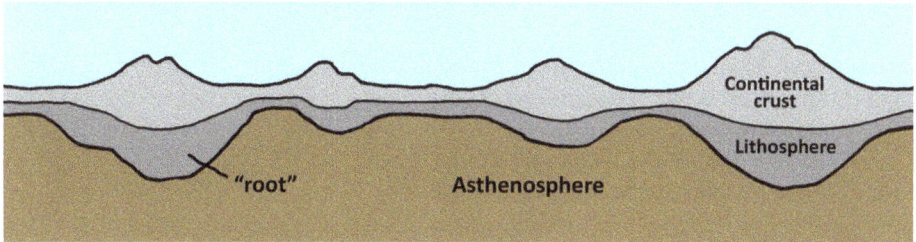

The interface between the lithosphere and the asthenosphere is anything but smooth. Even if the asthenosphere is somewhat plastic in consistency (and this is debatable) it is hard to imagine any way that the bumpy, lumpy continental crust would ever be able to move more than a tiny amount. It would take vertical thrust, not just lateral movement, to get a root out of its "hole." Also, getting one root dislodged wouldn't be enough. For the entire gray plate to move, ALL the roots would have to be dislodged at the same time. Plate tectonics has no answer for how this might happen.

Arthur Meyerhoff wrote this in 1996: "Today, geologists and geophysicists tend to treat pieces of the Earth's crust like a roomful of furniture, objects that can be pushed around at will into whatever configuration is required to satisfy a particular model. Unfortunately, the Earth's crust is not so easily manipulated...if one is faithful to physical laws." He states that the lithosphere is like a massive, 100-km deep interlocking mosaic of uneven rock, which makes plate motion extremely unlikely.[27]

Assumptions about the asthenosphere have also been called into question. Long-range seismic studies were done by Russian scientists during the 1980s and 1990s using underwater nuclear explosions and recording seismic waves in the rock beneath the seafloor to a depth of 100 km. These studies showed that the classical lithosphere-asthenosphere model is too simple. The asthenosphere is not one continuous layer all around the globe. It is much thicker in some places and is almost absent in other places. Additionally, seismic data clearly shows that the upper mantle is not homogeneous, but is stratified with alternating weak and rigid layers.[28] As Arthur Lerner-Lam of Columbia University remarked in a paper he published 1988: "Evidently, the earth has flunked the seismological test of the thin-plate theory." [29]

REASON #12: The rocks on the Mid-Atlantic Ridge should be the "youngest" rocks in the Atlantic, yet there is much data that challenges this claim.

Plate Tectonic theory says that new seafloor is created at mid-ocean ridges, so this is where we will find the youngest rocks (those with the lowest radiometric ages). The oldest rock will be in places very far from a spreading center.

As early as 1969, there were reports of continental rocks (such as granite) being found on the Mid-Atlantic Ridge. Most of the rocks dredged up from the Bald Mountain area just west of the Ridge at 45° N, were classified as continental rocks.[30] The scientists who examined the rocks commented that this was a "remarkable phenomenon," and began proposing theories to explain how the rocks got there. One idea was that they had been stuck in glaciers during the ice age and that the glaciers floated to this location, then melted, dropping the rocks on the ridge. Another idea was that these rocks had been used as ballast (weights) in sailing ships centuries earlier and that the ships had dumped their ballast in this location. However, estimates of the amount of loose rock on top of the Bald Mountain area came to about 80 cubic kilometers, so these ideas had to be tossed out. Further investigation of these rocks found them to have radiometric ages ranging from 169 to 1,500 million years.[31] According to plate tectonics, even the oldest parts of the seafloor should give radiometric dates of no more than 200 million years.

The USA's Deep Sea Drilling Program (DSDP), which ran from 1968 to 1983, investigated 624 sites, most of which were on, or near, ridges because those areas are shallower and therefore more accessible. They brought up core samples from as deep as 1,741 meters below the seafloor. The cores came up in 10 meter sections and were carefully labeled and preserved. The cores are now stored in three locations: in the US, Germany and Japan. Scientists can obtain small samples of the cores for research purposes. Although many descriptions of the DSDP proclaim that its greatest achievement was to provide deal-sealing evidence for Plate Tectonics, this statement is undeserved hype. In a report by petroleum geologist Arthur Meyerhoff, during "legs" 37 and 43 of the DSDP, igneous rocks considered to be Paleozoic and Proterozoic (250-2500 million radiometric years) were found in core samples from the Mid-Atlantic Ridge and the Bermuda Rise. Yet not one of these radiometrically ancient rocks was mentioned in the Cruise Site Reports.[32]

The ocean floor has three layers. The top layer is loose sediments and "ooze" up to .5 km deep. The second layer (1 to 2.5 km deep) is "flood basalt" from ancient magma flows. Thin sedimentary layers have been found mixed with the basalt. The third layer (averaging 5 km deep) is assumed to be gabbro, the "parent" rock that turns into basalt when melted. Beneath the gabbro is the upper mantle, presumably made of a rock called peridotite. No one has ever seen either the gabbro layer or the upper mantle. The drills did not even penetrate to the bottom of the flood basalts.

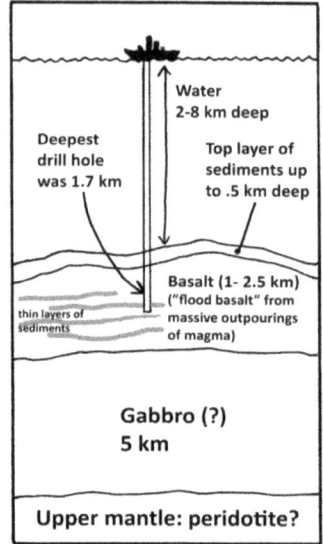

Water
2-8 km deep

Deepest drill hole was 1.7 km

Top layer of sediments up to .5 km deep

thin layers of sediments

Basalt (1- 2.5 km) ("flood basalt" from massive outpourings of magma)

Gabbro (?) 5 km

Upper mantle: peridotite?

Basaltic rocks having radiometric dates of 150-170 million years were found at the junction of the rift valley of the Mid-Atlantic Ridge and the Atlantis fracture zone (30°N), an area where basalt should theoretically be extremely young and have radiometric ages of less than 10 million years. Sometimes, the drills would bring up samples of basalt that gave radiometric ages younger than the rock on top of them, implying that basaltic magma had intruded into the area after the original rock had been formed. This added a new complication to the interpretation of the cores.[33] Also, cores could contain a layer of sediment *below* layers of basalt, a surprising find that no one had anticipated.[34] In many places, the drill bit would hit a layer of chert, a very hard silicon-based sedimentary rock, which would break the bit and cause that drilling site to be abandoned before they reached any basalt. No one seemed interested in why there was so much chert on top of the basalt. They simply removed the bit and moved on to another location.[35]

There is a group of tiny islands located right on the Mid-Atlantic Ridge. They were discovered by accident in 1511 by Portuguese sailing ships named *St. Peter* and *St. Paul*. The *St. Peter* ship ran aground on the rocks in the middle of the night and the sailors had to be rescued by the *St. Paul*. The islands are so small that they are often called the "rocks" of St. Peter and St. Paul. This is one of the very few places where the Mid-Atlantic Ridge is so high that it isn't under water, making it an ideal place to study ridge rocks. The rocks here, according to PT theory, should be less than 10 million radiometric years. However, rocks giving radiometric ages of up to 2,000 million years have been found.[36]

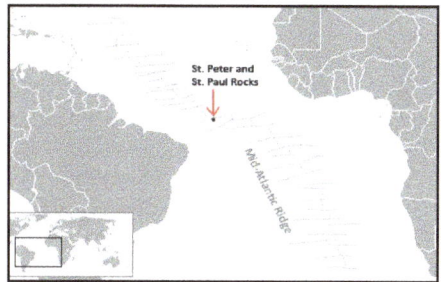

This equatorial region of the Mid-Atlantic Ridge has a surprising variety of rocks: quartzite, granitic metamorphic rocks, shale, brown coal and shallow-water fossils. Also, one seamount in this area is capped with 1,000 meters of sedimentary rock.[37]

The King's Trough lies just east of the MAR, in a region belived by PT theory to be relatively young, yet trilobite and graptolite fossils were found here. These fossils have traditionally been assigned radiometric ages of 320-540 million years, which is far older than PT predicts for any part of the seafloor.[38]

With all these "anomalies," why were drill cores touted as evidence that supported Plate Tectonics? The selection of data may have played a role. When the early drill core data (leg 3) was charted and graphed, and you ignored the fossil anomalies and the rocks brought up in dredge hauls (rocks from the surface of the seafloor), you could find a few places where the plotted points went in generally the right direction, with ages trending older going away from the ridge. Some have said that this trend was "cherry picked" for publication. There is a documented case of some geologists declaring that the date of a particular ridge rock specimen (635 million radiometric years) <u>must</u> be wrong because, as everyone knows, the rocks simply can't be older than 10 million in that location.[39]·

REASON #13: The "Bullard's Fit" doesn't fit.

On October 28, 1965, a paper entitled, "The fit of the continents around the Atlantic," was published in the Philosophical Transactions of the Royal Society of London. The paper had been written by Sir Edward Bullard, J. E. Everett and A. Gilbert Smith. They were not the first to notice the similarities of the coastlines on opposite sides of the Atlantic, but this paper marked the beginning of "the fit" becoming part of Plate Tectonic dogma.

The first glaring problem with "the fit" is the fact that Central America is missing. Central America has extensive areas of Paleozoic and Precambrian continental rocks, and the Caribbean Sea is underlain by ancient continental crust. The fact that a significant area of continental rock had to be trimmed out in order to make their idea work should have been a warning sign to Bullard, Everett and Smith that they needed to rethink their hypothesis.

A second problem is not as noticeable: they had to shrink Africa to get it to fit.[40] A third problem is visually obvious but its significance is not as obvious: the continents had to be rotated, some clockwise and others counterclockwise. How did seafloor spreading manage to not only move the continents apart, but rotate them? We don't see evidence of this rotational movement when we look at the ocean floor.

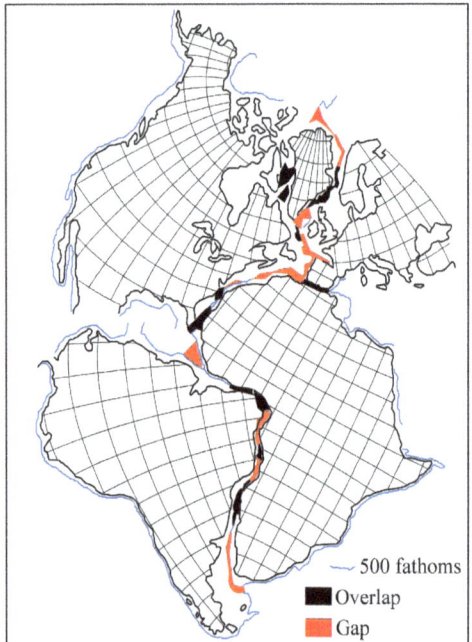

— 500 fathoms

■ Overlap

■ Gap

Jacques Kornprobst (redesigned after Bullard, E., Everett, J.E. and Smith, A.G., 1965. The fit of the continents around the Atlantic. Phil. Trans. Royal Soc., A 258, 1088, 41-51), CC BY-SA 4.0 <https://creativecommons.org/licenses/by-sa/4.0>, via Wikimedia Commons

One of the early objectors, E. N. Lyustikh, was very critical of the simple-minded shape matching of the coastlines of South America and Africa. He drew a diagram of fifteen coastlines around the world that have similar shapes, showing that the probability of any two coastlines matching is actually relatively high.[41]

Similarities of rock types and geological features along the coasts of South America and Africa are used to "prove" they were once connected. However, what is rarely mentioned are the dissimilarities. For example, western Africa and northern Brazil were supposedly once in contact, yet the structural trends of Africa run north to south, while those in Brazil run east to west. The Alps and the Himalayas have just as many similarities, but no one proposes they were once connected.[42]

As confirmation that something was amiss in the Bullard's Fit, other PT advocates have proposed alternative models. (Nafe and Drake, 1969, Dietz and Holden, 1970, Smith and Hallam, 1970, Tarling, 1971, Barron, Harrison and Hay, 1978, Smith, Hurley and Briden, 1981, Scotese, Gahagan, and Larson, 1988) Each of these alternative models takes liberties in cutting out parts of the continental crust in order to make their fit work.

REASON #14: Fracture zones along the ridges are said to show the direction of seafloor spreading in the past. As we map more of the ocean floor, we find many fracture zones and fault lines that challenge this hypothesis.

Fracture zones are the "cracks" that run perpendicular to the mid-ocean ridges. In addition to fracture zones, there are also faults (cracks). In this diagram, the Mid-Atlantic Ridge goes north to south and is shown by heavier lines. The fracture zones and faults are shown with lighter lines and go approximately east to west, (but with exceptions). Some of these lines have been highlighted in red and will be discussed below.

According to PT theory, fracture zones are very long cracks that were created by sections of seafloor as they spread at different times. The MAR is disjointed, with the ridge offset in some places by many kilometers. PT advocates say that the sections of

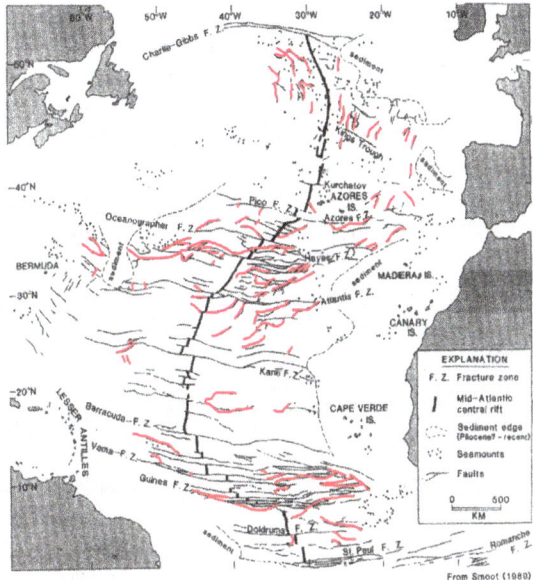

This map was drawn by bathymetry expert N. Christian Smoot, using the data he and his team members collected over many decades while on scientific cruises. The red lines were added by the author of this book to make clear which lines are problematic.

seafloor between the offsets would have spread at different times. They say that the fracture lines are where the crust cracked to allow this sideways motion. This explanation seemed to work in the 1960s, when only the largest ridges were known. Now, we know so much more about the topography of the Atlantic. Every red line in this diagram is a fracture zone or a fault line that flies in the face of the spreading hypothesis. Many lines in the north are parallel to the ridge, not perpendicular. Some of the lines form V or Y shapes. Some lines are curved. Some lines cross other lines. Every red line is an anomaly that does not conform to the theory.[43] These anomalies are never mentioned in textbooks; students are only shown fracture zones that align with the theory.

Please note that if you go to Google Earth to find these red line areas, you may or may not be able to find them. N. Christian Smoot, one of the world's most highly qualified experts on ocean floor bathymetry, says that Google Earth has relied far too heavily on satellite altimetry and in many cases did not bother to check with the more accurate data gathered by people like himself, who use sonar scanning devices attached to ships. The map shown here, Smoot says, is accurate. For ocean maps more accurate than Google Earth, Smoot recommends going to GEBCO.net.

REASON #15: Triple junctions are hard to explain, especially R-R-R junctions.

A triple junction is any point where three tectonic plates meet. The lines that meet at the junction can be ridges (R), trenches (T), or fault lines (F). The map below shows ridges in red, trenches in green, and transform faults in black. Junctions can be any combination of R, T and F, but the type we want to focus on is the R-R-R junction, where three red lines meet. The junction in the white circle will be discussed below.

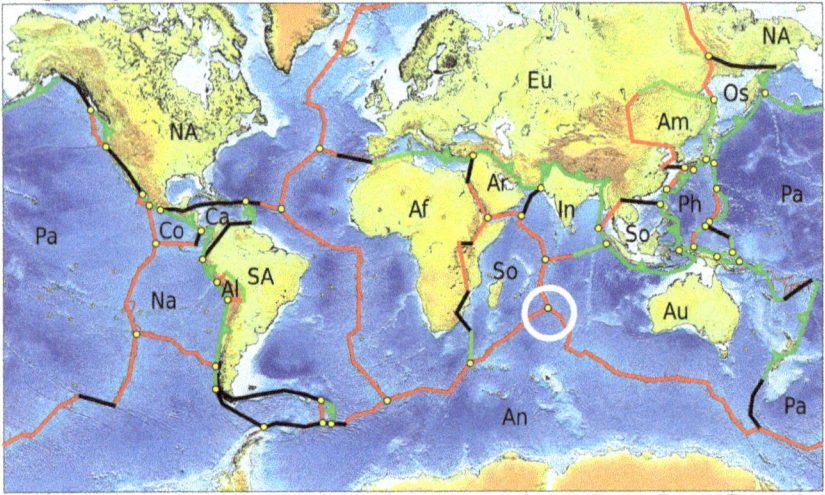

By Micheletb - Own work, CC BY-SA 4.0, https://commons.wikimedia.org/w/index.php?curid=76252790

The R-R-R (ridge-ridge-ridge) junctions are said to be places where a continent split apart, creating three divergent boundaries that radiated out from that central point. One of the boundaries "failed" and the other two went on to form an ocean. The definition of the R-R-R junction also includes the "fact" that the lines are at 120 degree angles.

One of the most notable R-R-R junctions, the Rodrigues Triple Junction in the Indian Ocean, is also one of the most problematic. It does not have 120 degree angles. On the right, we see a close-up of this junction. One of the angles is almost 180 degrees and appears to be one contin-uous ridge. This junction looks like the meeting of only two ridges, not three. The simplest and most obvious interpretation is that the continu-ous ridge (indicated by the line with an arrow at each end) formed first, then the other ridge ran

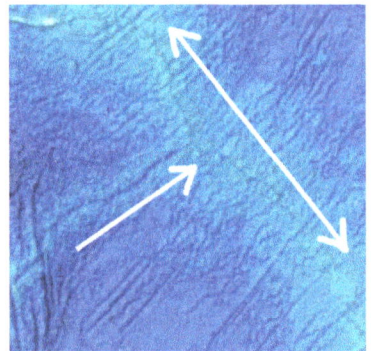

into it and stopped. This is exactly what we'd expect if the ridges are actually tension cracks, not spreading centers. Cracks can't "jump" across other cracks. If you try to tear a piece of paper in half both vertically and horizontally at the same time, the tear line that crosses the center line first will keep going to the other side. The "loser" will be stopped when it comes to the tear line.

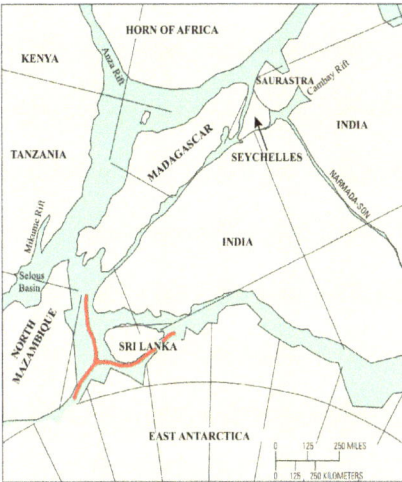

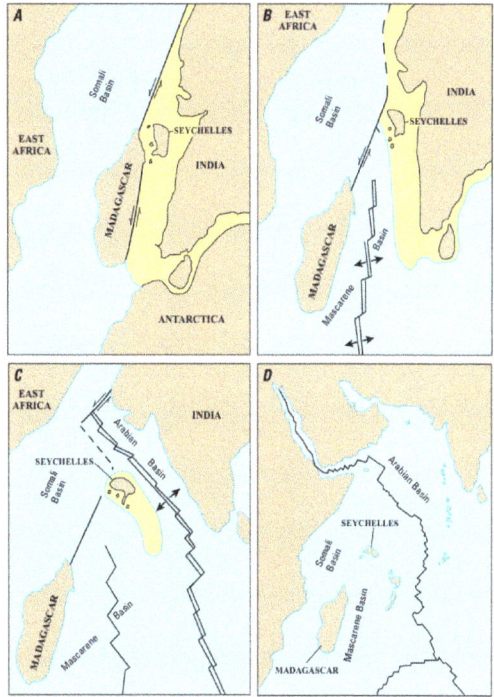

PT advocates, believing the junction to be part of seafloor spreading, have invented a complicated history, while at the same time admitting that they don't have much data to go on, and the history is "poorly understood."[44] The main type of data they use to study the junctions are measurements of magnetism on the ocean floor. The assumption is made that these areas of magnetism have recorded the direction of earth's magnetic field during past eras, and dates (of millions of years ago) are assigned to certain areas that have similar magnetism. Then a scenario is hypothesized by looking for patterns in the dates.

In the first map (top left), the future triple junction is shown in red. In the series of maps on the right, the conjectured history of this area is summarized. The Seychelles microplate is a main character in the story. The Seychelles islands are not extinct volcanoes like so many ocean islands, but contain continental granite, so they are assumed to have been part of a continent. A transform fault splits off Madagascar (A). Then a new spreading center opens up and Antarctica moves south (B). Then yet another spreading center opens between the Seychelles and India. The previous spreading center is shown with a single line, the new one with double lines (C). In the last diagram we see the triple junction after it has moved to the NE, away from Africa and Antarctica (D). Despite admitting the limitations caused by "the scarcity and complexity of identified magnetic delineations," the foremost researcher on this junction goes on to give quite precise details about the rate of its NE movement: 3.6 to 3.8 cm/yr from present to 41 million years ago, 2.6 cm/yr from 43 to 65 million years ago, and 10 cm/yr before 65 million years ago.[45]

The process by which all this happens is assumed to be "upwellings of magma from the mantle." The mechanisms for upwelling are left for other branches of the geophysics community to investigate.

Let's look again at the close up view of this junction to see another problem. Do we see any evidence along these divergent zones of ridges from bygone eras, as the spreading happened bit by bit, over time? Image A shows the ridges as they are. We see only one central crack along the ridge. In the SE portion of the map, we hardly see the ridge at all, but there are extremely long perpendicular fracture zones going out from the ridge.

Image B has been Photoshopped® to be closer to what we would expect to see. If the seafloor has been spreading for as long as they say it has, and if spreading is initiated by the opening of a new fissure along the ridge, why don't we see all these past ridges? Shouldn't there be a series of "dead" ridges parallel to the current active ridge? Image B should also show mountainous areas in the triangular regions between the ridges, as this area is subject to spreading from two direc- tions at once and therefore would have double the volume of seafloor rock. In fact, these regions are very flat.

Other triple junctions have these same issues. The Galapagos Triple Junction is shown in the bottom map. The vertical line on the left is part of the East Pacific Rise. Notice that again we do not see three 120 degree an- gles, but a T shape. The EPR line lacks fractures zones, and has no parallel ridges. Where is the evidence of spreading? The Bouvet Triple Junc- tion in the South Atlantic is also a T, and lacks parallel ridges.

Histories have been invented for the dozens of triple junctions around the globe. For some junctions, their history is said to be "lost in time" as parts of them have already subducted under continents.

If these ridges are not spreading centers, but are the remains of tension cracks in the mantle (tension in two directions, one creating fracture zones, the other creating central ridges) there is no need to invent complicated histories for these junctions. The observed features fit the tension crack hypothesis very well. The newer crack stopped when it ran into the older crack.

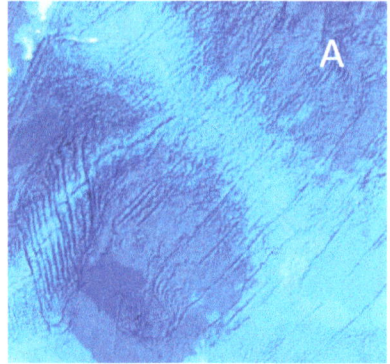

Both A and B show the Rodrigues Triple Junction in the Indian Ocean.

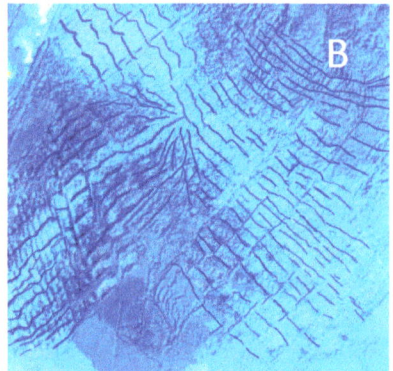

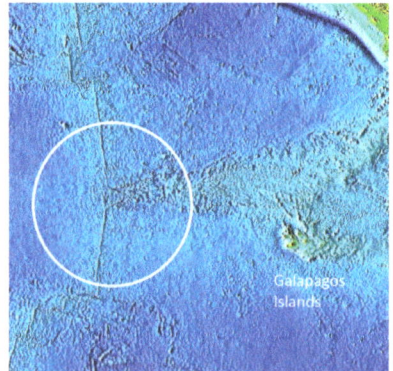

The Galapagos Triple Junction

REASON #16:   Plate Tectonics doesn't take into account "megatrends."

A megatrend is a continuous linear formation made of any combination of features: ridges, seamount chains, fracture zones, or fault lines. Megatrends were discovered and named by N. Christian Smoot in 1994. Smoot was a senior scientist for the Ocean Survey Program of the US Naval Oceanographic Office. His expertise was multibeam sonar, but late in his career he also began looking at satellite altimetry images. (Satellites can measure the surface of the ocean very accurately; because of gravity, the water level at the surface reflects the shape of the ocean floor beneath it.) When he combined his bathymetry images with the satellite images, he discovered distinct linear patterns. In general, the lines fanned out from the trench zones, splaying eastward. Many lines made sharp V's or made orthogonal (square-ish) intersections.

The megatrends, which incorporate all the fracture zones, cross trenches and ridges. Some even cross the entire ocean (red line). They pass through rocks that are radiometrically older than the supposed ages of the oceans themselves. Some mega-trends keep going right into continents. The Mendocino Fracture Zone (orange) contin-ues into North America, going all the way to Yellowstone. The Mendocino also crosses the San Andreas Fault line (green) but does not show evidence of the shifting fault. The lines highlighted in purple show examples of megatrends that either have orthogonal in-tersections, or create fan-shape patterns, both of which prove that fracture zones can't possibly be a record of past seafloor spreading, as PT theorists claim.

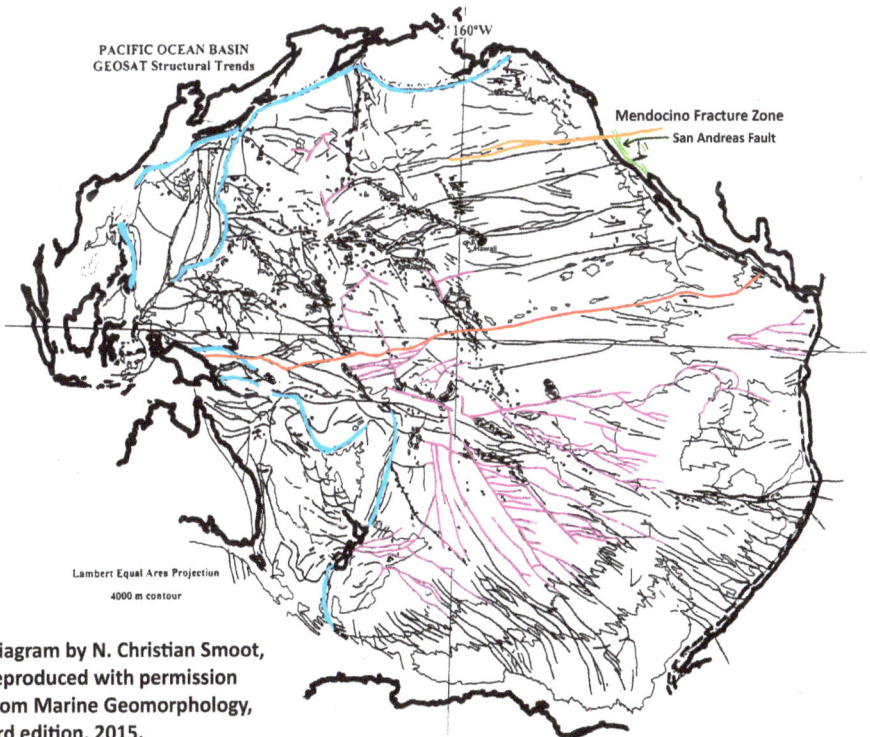

PACIFIC OCEAN BASIN
GEOSAT Structural Trends

160°W

Mendocino Fracture Zone
San Andreas Fault

Lambert Equal Area Projection
4000 m contour

Diagram by N. Christian Smoot,
reproduced with permission
from Marine Geomorphology,
3rd edition, 2015.

REASON #17:   Magma is a compressible fluid and sinks when below 200 km. This means that mantle convection and mantle plumes are impossible.

The hypothesis of seafloor spreading is directly connected to another hypothesized process: mantle convection. As shown in this diagram, PT advocates suggest that the mantle slowly circulates in convection cells. Textbooks often compare this to the circulation of boiling water in a pot. Seismic studies have repeatedly shown, however, that the mantle is solid. Therefore, geologists claim that although the mantle is solid, it is able to circulate "on geological timescales." In other words, they can have it both ways as long as they invoke vast amounts of time.

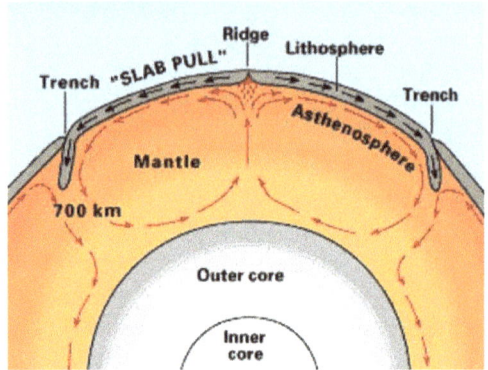

One of the very few experiments that has shed some light on the question of mantle convection was done in 1998 at Okayama University in Japan.[46] The scientists put a small piece of basaltic rock into a device called a diamond anvil chamber and then turned up the heat and pressure to levels believed to be found deep in the mantle. They found that after the rock melted into basaltic magma, increasing the pressure caused the liquid magma to compress, taking up 50% less space, thereby greatly increasing its density. The final conclusion that the researchers came to was that magma produced below depths of about 200 km would be more dense than the surrounding rock, and would thus be unable to rise. (The depth at which this happens is called the "cross-over depth.") Previous experiments that used techniques other than the diamond anvil had suggested that basaltic magma below about 300-400 km would be unable to rise.[47, 48, 49] Even if the cross-over depth is as great as 400 km, this still makes the idea of mantle convection unrealistic.

This diagram shows how deep 200 km is compared to the entire mantle. Below the purple line, magma sinks and cannot rise to the surface. Volcanism must be essentially a surface phenomenon, not the result of whatever is happening deep in the mantle. Any melting that occurs under the red line would drain downward, perhaps eventually ending up in the liquid outer core.

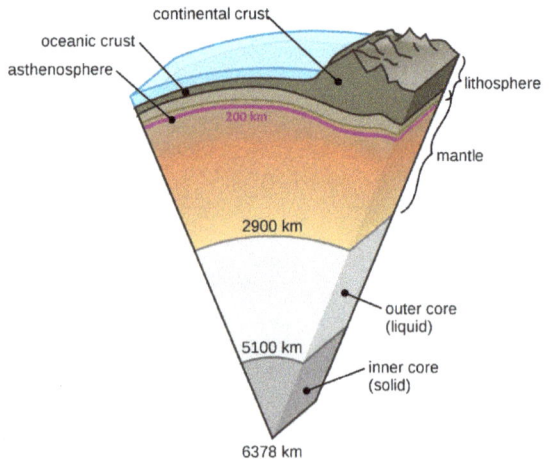

If the mantle can't convect, that means that the idea of "mantle plumes" must also be impossible. Plumes are areas were magma is thought to rise from earth's outer core. When a plume finally reaches the surface, it can start or maintain seafloor spreading, break apart continents, or create a volcano either on land or in the ocean. Plumes are necessary to explain volcanoes that are far from convergent zones.

According to the Wikipedia article on mantle plumes (using Wikipedia as a gauge of current mainstream thought in geology) the key evidences for plumes are:
1) Seismic images
2) Linear chains of volcanic islands (where the plumes are called "hotspots")
3) Computer models that show that it's true

Setting aside the fact that we just learned about an experiment that showed that magma can't rise when it is below 200 km, there are other problems with these evidences.

1) Seismic images are open to more than one interpretation, as we saw in Reason #7. Seismic data only tells us how fast P and S waves travel through that area. Multiple factors contribute to how a rock behaves. We can't assume that just because an area transmits waves more slowly that we can know for sure that the area contains a certain type of magma. It's a possibility, but not a surety. The super deep boreholes served as a warning to geologists not to be too confident about their interpretations of seismic data. The rock that came up from the holes did not confirm their predictions.

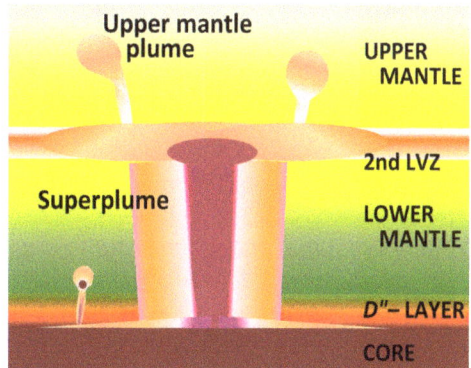

Mantle plume theory has become quite complex over the years. Up to 1,500 plumes have been hypothesized, compared to the original 7 proposed in the 1960s.

By Brews ohare - Own work based upon Matyska & Yuen (2007) "Figure 17 in Lower-mantle material properties and convection models of multiscale plumes" in Plates, plumes, and planetary processes, Geological Society of America, p. 159 ISBN: 0813724309., CC BY-SA 3.0, https://commons.wikimedia.org/w/index.php?curid=9454468

A tomography image is a snapshot, not a video. The assumption is made that the still tomography image shows an area of *rising* magma, but it could just as easily show an area where magma is draining *downward*, into the core. Even if a tomography video could be taken, they'd still not expect to see any movement, as they think that the plume rises over eons of time. So believing that the image shows sinking magma is just as valid as believing that it shows rising magma. Considering that experiments have shown that magma becomes more dense than the surrounding rock at a depth of about 200 km (as discussed on the previous page), the chance that the image shows magma going down, not up, is fairly high.

2) The "hot spot" theory says that when a tectonic plate passes over a place where a plume has reached the lithosphere, magma can break through from time to time, creating a chain of volcanoes, some of which might become volcanic islands.

The hot spot hypothesis was created by the original PT theorists in order to try to explain the anomalous volcanoes that are not near a plate boundary, as the theory says they should be. They obviously had not completely thought it through before they proposed it. Here are some issues to consider:

A) If the mantle circulates enough to move the plates, how can there be a fixed plume in a moving mantle?[50]

B) If a chain of volcanoes means the plate is drifting, does an isolated volcano mean the plate is not drifting? (The Pacific has over 40,000 randomly scattered volcanoes.)

C) Some volcanic chains, such as the Bermuda Rise, are almost perpendicular to the claimed movement of their plates. Also, the Hawaiian Islands are part of a larger chain that includes the Emperor sea mounts. This mega-chain has a bend of 60° in the middle. What created the bend?

D) Faster moving plates should have fewer volcanic cones "burned" through them than slower moving plates. It seems that the opposite is the case. The Pacific plate is said to be moving the fastest and it has by far the most volcanoes.

E) Heat transfer equations can show that if the earth is billions of years old, as they claim, by now temperatures of the core and mantle would have evened out, ruling out areas like plumes that are much hotter than other areas.

F) Pressure in the lower mantle is so high that it will not allow cracks to open up so that magma can rise.

G) If plumes rise slowly, as they claim, heat transfer equations can show that the heat of the plume will dissipate into the surrounding rock long before it reaches the lithosphere.[51]

H) It has been claimed that there is a radiometric age progression along the chain of the Hawaiian Islands (though this has been disputed by other scientists), but other chains of islands definitely do not show this type of progression.[52]

I) The chemistry of the rocks comprising these hot-spot islands suggests that the magma came from the upper mantle, not the lower mantle. This bit of data has caused some geologists to rethink the mantle plume idea and suggest instead that most magma is coming from a relatively shallow source, not from deep in the mantle.

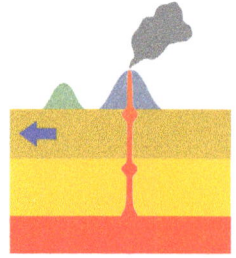

3) Computer models can be designed to show anything. You can control the inputs in order to achieve the desired outputs. This illustration shows a sample of a computer simulation of mantle plumes.

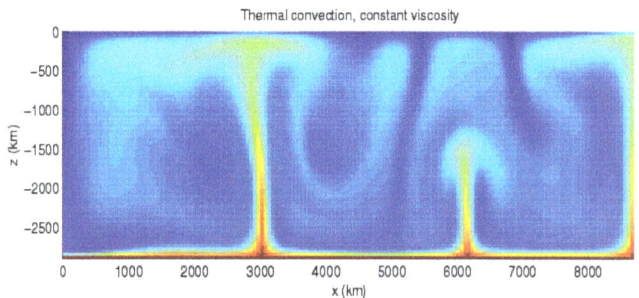

Thermal convection, constant viscosity

Many of the arguments against mantle plumes and hotspots seem so obvious and so valid that you may be wondering why geologists adhere so tightly to the ideas. The answer comes from Warren Hamilton, a distinguished senior scientist in the geophysics department of the Colorado School of Mines. He wrote more than 100 peer-reviewed papers, spent 43 years in the U.S. Geological Survey, and received the highly valued Penrose Medal. He was a plate tectonics advocate, but late in his career he saw that mantle plumes were not possible, and began publishing other ideas that involved processes in the lithosphere and upper mantle. He soon ran into fierce opposition and in some cases downright hostility. His frustration finally culminated in an essay in which he openly accused the scientific establishment of being narrow-minded and anti-progress.

"Different rules of evidence apply to current paradigms than to challengers. For current paradigms, evaluation is not necessary; for challengers, proof must be overwhelming before even a discussion is warranted. Appalling papers in support of mantle plumes are now being published in major journals, perhaps indicating that pro-plume reviewers will welcome any debris that will heighten the rampart against a paradigm shift.

Suppression of dissent is common. I know of many examples of prominent members of the pro-plume community stifling anti-plume reports and research. Although these stiflers might tell themselves that they are maintaining professional standards, "bad science" to them means anything contrary to their own beliefs. Many reviewers block grants to, and prevent publication by, anyone who holds a contrary view.

Widely accepted assumptions regarding the composition and behavior of the mantle appear to be in error, yet evaluation of alternatives is actively discouraged. The present enthusiasm for plumes represents groupthink that is easily falsified and yet is impervious to evidence. The fact that we are scientists confers no infallibility upon us, and our egos often lead us astray." [53]

Another notable anti-plume geophysicist is H. C. Sheth, of the Physical Research Lab in Navrangpura, Ahmedebad, India. He is an expert in flood basalts on land, especially the Deccan basalts which cover a vast area of India. Most geologists would say that the Deccan flood basalts came from a mantle plume. Sheth disagrees. He says, " Hypothesized mantle plumes do not appear responsible for most large igneous provinces; instead, their very existence is questionable. No geological evidence of any kind—geochemical, petrological, thermal, topographic—requires mantle plumes."[54]

Gillian Foulger of the University of Durham, UK, saw the same "Procrustean bed" problem that the Meyerhoffs had pointed out in the early years of the theory. Mantle plumes can apparently have any characteristics you want to attribute to them. She says: "Plumes have been proposed to come from almost any depth, to rise vertically or tilt, to flow for long distances laterally, to have narrow or broad conduits, to have no plume head, one head, or multiple heads, to produce steady or variable flow, to be long or short-lived, to speed up or slow down, to have a source that is either depleted, enriched, or both, and to have either high or low ratios of helium-3 to helium-4."[55]

Wikipedia does now list two alternatives to mantle plume theory: the "plate hypothesis" and the "impact hypothesis." The plate hypothesis was put forward by Warren Hamilton, Don Anderson, and Gillian Foulger. They attribute volcanism to shallow, near-surface processes. The impact hypothesis proposes that impacts from meteors resulted in short periods of extreme volcanism.

REASON #18:   The deepest holes ever drilled (Kola peninsula, Germany) showed some plate tectonic predictions to be wrong.

The Kola Superdeep Borehole is the deepest hole anyone has ever drilled. It was drilled by Russian scientists from 1970 to 1994, near Murmansk, in the Kola peninsula, to the east of northern Finland. The Kola derrick, shown on the right, was a large structure compared to the bore-hole itself, which was only about 23 centimeters in diameter. The top of the hole was welded shut (bottom photo) after the project ended in 1994. Today, the site is an abandoned ghost town. There were actually five holes drilled at the site, but hole #3 was the record-breaker at 12,262 meters (40,230 ft; 7.619 miles). They had to stop drilling at 12,262 meters because every time they brought the drill up to put on a new head, the hole filled back in and they had to redrill it. The redrilling would dull the bit, so they'd have to pull it up again before they had drilled any deeper. After several repeats of this cycle, they gave up.[56]

The goal was to drill all the way through the continental shield rock and into the mantle. They only got one third of the way through the granite shield. Based on their seismic studies, geologists expected a transition from granite to basalt at about 7 kilometers, but instead they found that at this level the granite had undergone a metamorphic transition, changing its density. No basalt was found. A little deeper they found something no one had predicted. The granite had been thoroughly fractured and the fractures were filled with salty water. It seemed impossible for water to be at this depth. It can't possibly have leaked down from the surface because fresh cracks would close up due to pressure. It must have somehow come from below.

Another big surprise was found at 6 kilometers deep: marine plankton fossils. Also, equally surprising, was the large volume of hydrogen gas that came up. The drilling mud that came out of the hole was described as "boiling" with hydrogen bubbles.[57]

The rate at which the heat increased with depth was also out of line with plate tectonic predictions. The temperature got hotter much faster than expected. At the final depth of 12.3 km, the expected temperature was 100 °C (212 °F); what they found was 180 °C (356 °F). Plate tectonics assumes that heat increases steadily with depth. If the temperature was to keep increasing at the rate that was found in the borehole, the calculated temperature of the inner core would have been reached at a shockingly shallow depth.

The big take-aways from the Kola borehole were: 1) Interpretations of seismic data can be very wrong, and 2) it is incredibly difficult to drill down to the mantle and it is therefore unlikely this will become a readily available source of data.

In 1987, Germany decided to drill its own superdeep borehole (the KTB) in a location that they were sure was right on the boundary of a former (hypothesized) continental plate that has since disappeared. They wanted to go right through that zone and bring up rock samples as evidence of the assumed ancient collision. The site chosen was near Windischeschenbach, Bavaria. First, they drilled an experimental pilot hole, to test their equipment. It went down to 4,000 meters. The German scientists had studied the problems that the Kola drillers had run into and had tried to design a new type of drill head that would avoid those problems. They also devised a way to make the drillhead self-correcting when it sensed that it was going off-plumb. The German holes ended up being much straighter than any other boreholes ever drilled, either before or after. The second hole was started in 1990, about 200 meters away from the pilot hole, and it reached a depth of 9,101 meters.

What did the German scientists expect to find, working from their knowledge of plate tectonics? They predicted that they would find evidence of an ancient nappe—a large, flat area of rock that had been pushed over a lower layer—from 3 to 5 kilometers thick. Below that they expected to find the "suture zone" where a collision had taken place, and where there should be an abundance of metamorphic rock that would reflect all the stress this area had been under during past eons.[58]

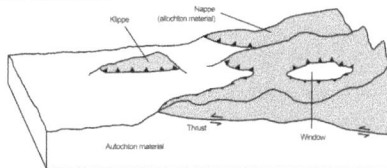

Gray area is a "nappe." None was found.

The nappe was never found. Instead, they found steeply inclined layers of rock, as if they had been folded and then pushed together. Indeed, these layers did contain a lot of metamorphic rock, as predicted, though not the expected types and not the expected texture. The parent rock that these layers came from seemed to be thin lava flows (chemically similar to mid-ocean ridge basalt) and limestones. The layers were steeply angled all the way to the bottom so the drill never reached a suture zone.[59]

As with Kola, the biggest surprise was the amount of crushed rock and water, even very salty water, at great depths, along with considerable amounts of gases such as methane and nitrogen. The chemical signature of the nitrogen and methane led scientists to believe that these had come from the decomposition of organic material in a marine sedimentary environment. Near the bottom of the hole, the salty water was twice as salty as ocean water and contained not only sodium and chlorine, but also calcium and strontium. Also, surprising amounts of graphite (the stuff that pencil "leads" are made of) were found in thin layers here and there. Graphite is very slippery (which is why your pencil works) and is perhaps an under-appreciated factor in de-termining why faults (slipping rocks) occur where they do. Graphite also has interest-ing electrical properties. The KTB confirmed the temperature data from Kola.[60] The temperatures went up much more quickly than anyone expected, suggesting that the current theories about the sources of geothermal heat are incorrect.

The German KTB borehole taught geologists the same lesson that the Kola borehole did: 1) seismic studies give us very limited information about what is deep underground, and 2) plate tectonic theory has not proven useful in predicting what we'll find when we dig deep into the ground.

REASON #19: If the earth has been recycling its plates in and out of the mantle for billions of years, shouldn't the lithosphere and mantle be more homogeneous by now?

We don't know for sure what the mantle is made of because we've never been able to drill down into it and bring up a sample. This point can't be emphasized enough. The only *facts* (direct observations) we have about the mantle are measurements of how fast seismic waves travel at certain depths. The superdeep boreholes showed that geologists can be very wrong in their interpretations of what those seismic velocities mean. However, we can make educated guesses about the mantle by looking at xenoliths—chunks of unmelted rock that come up with magma. We assume that the xenoliths are pieces of mantle rock that had not yet melted when they were swept up in the flow of magma and brought to the surface.

A piece of basalt with a light green xenolith made of olivine.

In basaltic rocks, xenoliths are usually made of the silicon-based minerals olivine and pyroxene, or a mixture of both called peridotite. The xenoliths are easy to spot, as they often appear as bright green crystals against a dull gray background. Occasionally, red crystals of garnet or spinel are found in xenoliths.[61]

When peridotite, olivine and pyroxene melt, they turn into runny lava that cools into either fine-grained basalt or coarse-grained gabbro. We know this because we can heat these minerals in a lab and watch what happens. We also know that when basalt is heated it doesn't turn back into olivine or peridotite. It's a one-way transformation. You can watch videos on YouTube of people melting volcanic rocks in homemade furnaces. The lava they make cools into hard, black lava rocks.

Most geology texts will tell you that the mantle is made of olivine, pyroxene and peridotite, citing xenoliths as proof. However, if mantle convection has been going on long enough to have recycled entire oceans and continents, wouldn't we expect all of the mantle's source rocks to have melted into basalt or gabbro? Surely in all that time, wouldn't a few complete cycles of those convection arrows have occured, with all of the rock having been melted at mid-ocean ridges at least once?

An ancillary problem is to consider whether it is possible for the mantle to make continental crust. Wikipedia (being a repository for mainstream thought) says the following: "Partial melting of the mantle at mid-ocean ridges produces oceanic crust, and partial melting of the mantle at subduction zones produces continental crust." Is this possible? The bottom layer of continental crust is primarily granite, not basalt. Both types of rock are silicon-based rocks, but granite is made of minerals such as quartz, feldspar and mica, and contains very little magnesium and iron, whereas basalt has much less quartz and has a lot of magnesium and iron. The article never gets around to telling you exactly how the mantle is able to make both basalt and granite.

granite

REASON #20: Even though magnetic anomalies helped PT theory to gain acceptance, there are many problems surrounding them.

Magnetic anomalies on the seafloor were discovered in the 1950s. During World War II, fluxgate magnetometers had been developed for finding submarines hiding deep in the ocean. After the war, they continued to be used for finding metal objects over both land and sea. Then, someone had an interesting idea—what would happen if you dragged a magnetometer behind a ship that was crossing the ocean and had it record any changes in magnetism that it could detect on the seafloor? The results were interesting enough that scientists were able to obtain grants to charter ships that would sail back and forth over various locations and gather a large amount of data.

One of the first locations to be studied was a portion of the Mid-Atlantic Ridge just south of Iceland, called Reykjanes *(Ray-kyah-ness)* Ridge. The magnetometer recorded fluctuations in the strength of the magnetic field as it crossed the ridge. When the data was analyzed, a striped pattern emerged. Since the areas along the mid-ocean ridge are more shallow than other parts of the ocean, the scientists were able to do some drilling in these areas to obtain core samples from both sides of the ridge. When it was found that there was some symmetry to the radiometric ages, and that the radiometric ages appeared to increase with distance from the ridge, this was hailed as "proof" that the theory of seafloor spreading was correct. A few geologists warned that no one should jump to conclusions before more data was gathered, but jump they did. Anyone who didn't go along with the jump was ostracized. The Reykjanes magnetic stripes have become icons of plate tectonic theory, and are often the only magnetic data a student ever sees.

An artist's facsimile of the magnetic stripe pattern along Reykjanes Ridge. Similar gray tones means similar radiometric ages.

The first person to suggest that there was a connection between seafloor spreading and magnetic anomalies was Canadian geologist Lawrence Morely, with his letters to *Nature* in the spring of 1963. After rejection of Morley's proposals for publication, the editors at *Nature* then decided to publish these same ideas in September of 1963, when they were submitted by British scientists Frederick Vine and Drummond Matthews. The proposal was at first called the Vine-Matthews hypothesis, but as time went on and it became known that Morley had been the first to write to *Nature*, the official name of the hypothesis was changed to "Morley-Vine-Matthews."

The hypothesis went like this: When magma rises at the ridge, it is so hot that all magnetism in its minerals disappears. (The temperature at which materials like iron lose their magnetism is called the Curie point.) The magnetic minerals in the hot magma are free to move around, and, while free to move, they line up with the earth's magnetic

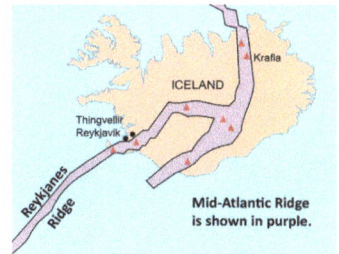

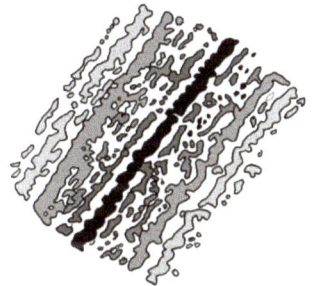

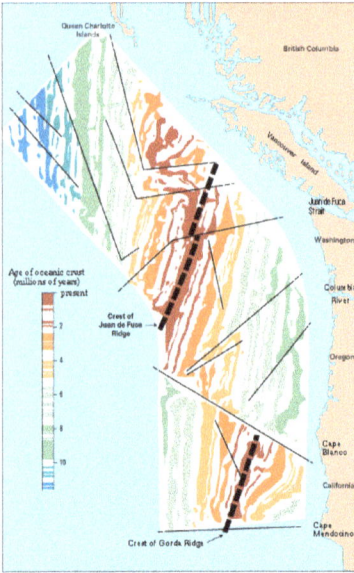

field, as if each atom was a little floating compass. As the magma cools into a solid (rock), the temperature falls below the Curie point, and the mineral crystals get locked into place. They will stay locked in place as long as the temperature stays below the Curie point. In this way, rocks could act as a historical record of the earth's magnetic past.

The negative magnetic anomalies (areas of low magnetic intensity) recorded by the magnetometers were interpreted as areas of reversed magnetic polarity—times when the earth's north and south poles were reversed.

These stripes along ocean ridges seemed to be watertight proof that seafloor spreading had, indeed, occurred, and at a rate that could be determined by radiometric dating.

Above is another famous image from the early days of magnetic anomaly research. This image first appeared in Scientific American *magazine in 1961. Because the article pre-dates the Morley-Vine-Matthews hypothesis, it gives several ideas for how the data might be interpreted.*

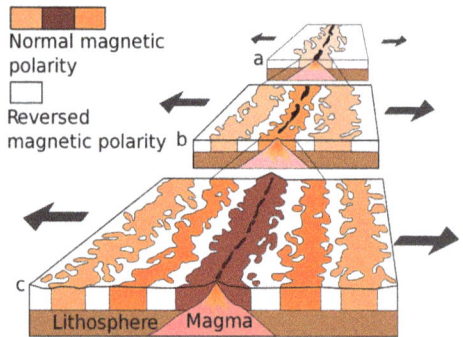

In the late 1960s, other areas of the ocean were studied using magnetometers. It turns out that the first places to be mapped just happened to be ideal for the seafloor spreading hypothesis. Other areas are ambiguous at best.

Here we see an artist's facsimile of data from 1966, showing the North Atlantic to the east of the Mid-Atlantic Ridge. (Data was not gathered directly north of Iceland because the MAR disappears in this area, presumably buried by massive lava flows.) It would have been hard to formulate the M-V-M hypothesis using the anomalies from this area. We do see somewhat of a striping pattern in certain areas, but also a complete lack of striping in others.

In 2009, the USGS published the map you see below. Red areas are positive anomalies (assumed to be areas of normal magnetic polarity) and blue areas are negative anomalies (assumed to be areas of reversed magnetic polarity). You can see the Reykjanes Ridge area at the upper right (Iceland is white), and the north-eastern Pacific area on the left, off the western coast of British Columbia, Washington and Oregon.

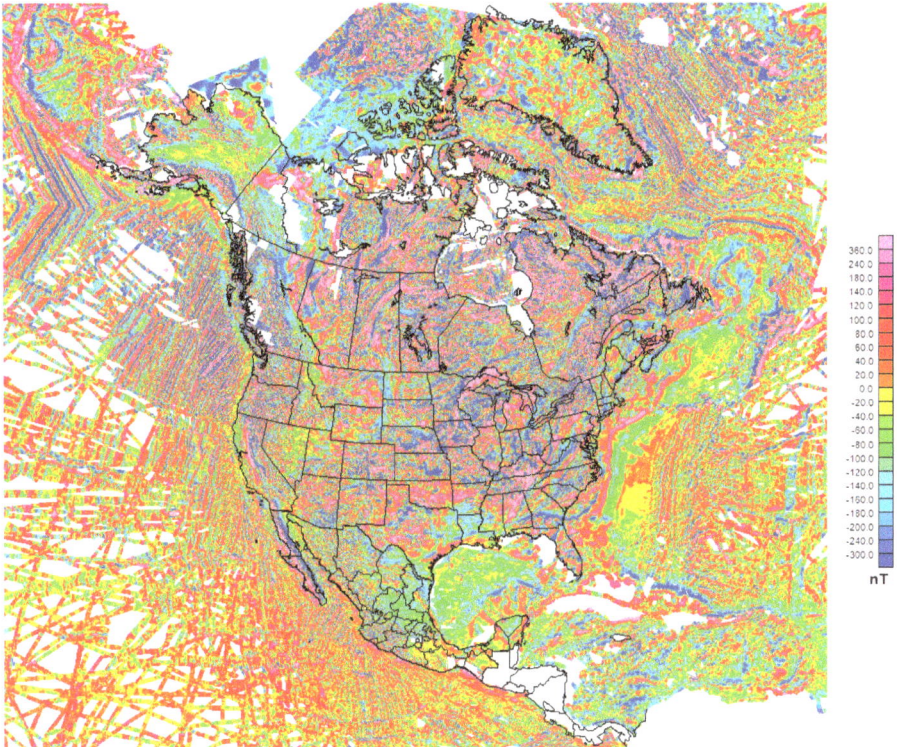

It is very easy to see that many parts of the ocean can't be used to support the M-V-M hypothesis. How could seafloor spreading have created the V-shaped lines in the upper left corner? Why can't we easily see the East Pacific Rise (supposedly a spreading site)? And what about the magnetic patterns on land? What do those mean? Could rock type have anything to do with the patterns? The western coast of north America has many rocks that contain the mineral magnetite. The idea that the patterns were related to the susceptibility of certain rocks to magnetism was proposed by a number of researchers (Mason and Raff, 1961; Drake and Girdler, 1964; Einarsson, 1967; Gudmundsson, 1967; Luyendyk and Melson, 1967; Harrison, 1968; van Andel, 1968). Luyendyk and Melson wrote that these striped zones might be "related mainly to zones of fresh basalts, alternating with hydrothermally altered zones which perhaps surround fracture zones." [62] The research done by Luyendyk and Melson on the rock types in these famous areas indicated that their magnetic susceptibility was probably capable of producing the anomaly patterns. However, once the M-V-M hypothesis was generally accepted, other hypotheses were either abandoned or ignored.

The M-V-M interpretation of magnetic anomalies fails to take into account a number of factors. First, a significant number of ocean floor rocks did not come from cooling magma. Serpentinite, for example, appears to be mantle rock (peridotite) that absorbed seawater. This reaction happens at temperatures between 125° and 500° C. The Curie point for magnetite (the primary and most important magnetic mineral found in mantle rock) is 580° C. Therefore, any place we find serpentinite, no cooling through the Curie point has occurred. Serpentinite is found on mid-ocean ridges, which are said to be spreading centers created by magma coming up from the mantle below. Ridges also contain a significant

The serpentinite on the left has obvious streaks of black magnetite. The serpentinite on the right is a homogeneous mix of magnetite with other minerals.

amount of gabbro, another mantle rock altered by low temperatures. The presence of serpentinite and gabbro along ridges challenges not only the M-V-M hypothesis, but plate tectonic theory, as well. Serpentinite is also found in great abundance along the western coast of North America. It was the waters off this coast that provided the first map of magnetic anomalies. (pg. 52). If the seafloor in this area contains a significant amount of serpentinite, the M-V-M narrative about cooling magma must be re-evaluated.

Second, magnetometers can't accurately pinpoint the depth from which the magnetism comes. We don't know for sure if the normal and reversed polarity readings are from lava rocks close to the surface, or from deeper rocks. In an attempt to shed light on this question, researchers in 1998 used a submersible vehicle to scan the face of an underwater cliff (Blanco Scarp in the NE Pacific) that was 3.5 km high. They found that polarity reversals occur at various depths, down to 3.5 km.[63] This confirms previous research done in 1979, when deep sea drillers found reversals occurring at various depths.[64] To explain this unexpected depth data, PT theorists had to speculate that lava is extruded not only in the center of a spreading ridge, but along its sides as well. Geological observations of the ridges, however, don't support this idea.

Third, there is a strong correlation between magnetic anomaly data and the topographic relief of the ocean floor (underwater mountains and valleys). Though efforts are made to compensate for the effects of topography, there is a wide range of opinions.[65]

Fourth, a variety of factors might come into play after the original magnetization of the rock. Re-magnetization can occur by either thermal or chemical processes. Re-heating of the rocks can allow the magnetism to change. Chemistry can also play a role, as magnetite crystals can align themselves with magnetic fields during the natural process of crystal growth.[66] Chemistry can also cause magnetism to be lost. Additionally, the grain size of the minerals in the rock can determine how susceptible a rock is to magnetization. Magnetite grains can't be permanently magnetized unless the crystals are single domain with grain sizes less than 10 microns.[67]

large magnetite crystals

A fifth difficulty is with the assumption that magnetic reversals took place over very long spans of time. The graph shown here is a facsimile of raw data from a magnetometer. Lines are drawn at places where the intensity falls below average, and these are assumed to indicate periods of reversal. The black bars show periods ("chrons") of normal polarity and the white bars are chrons of reversed polarity. The chrons can be short or long, and seem chaotic, with no regular pattern. The time it takes for the polarity to reverse is said to be centuries or millennia. This belief has been challenged by research done on a dyke formation in the Linzhou basin (China), near the area believed to be the "collision" zone of Asia and India. This dyke is a "sheet" of rock formed by magma that rose quickly through several-meter-wide cracks in the continental rock. We know this event had to occur rather quickly because once magma cools, it will block the flow of more magma from below. To form a dyke reaching to the surface, the magma has to rise the entire distance pretty much in one go, probably within hours. In this dyke, both normal and reversed polarity were detected. The scientists were careful to rule out the phenomenon of self-reversing magnetism that can occur rarely in some rocks. They were sure that these reversals were caused by the orientation of the earth's magnetic field at the time the magma cooled.[68] This suggests the possibility that the earth's polarity can reverse quickly, maybe even instantly, but more research is needed to confirm this idea. We know for a fact that the sun's polarity reverses when the sunspot cycle reaches its peak, about every eleven years. The sun is very different from the earth, of course, but it does show us that when conditions are right, reversals can be fast.

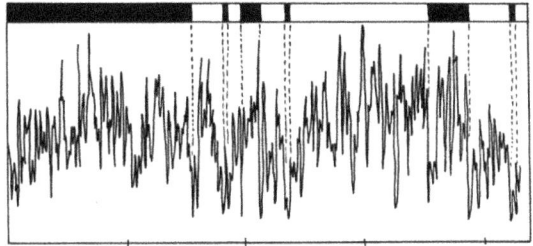

Problem number six is that plates have to be invented in order to explain some of the magnetic striping patterns. For example, in an area called the Great Magnetic Bight, off the coast of Alaska, there is an odd V-shaped pattern (upper left corner in the map on page 53). It has been proposed that this pattern is the remnant of a triple junction and a spreading center that started in a more central Pacific location then migrated to the northeast (despite the fact that the Pacific plate now moves northwest), with the spreading center and two of the plates either partially or completely disappearing as they subducted under continental crust. This invented history is so complicated that it can take several pages to explain.[69] We have no actual proof that these plates or this spreading center ever existed, only the opinions of geologists who insist that their history must be correct because, after all, we see this odd pattern of magnetic striping.

Problem number seven deals with statistics. Several seasoned researchers decided to take a closer look at the magnetic data collected on both sides of the Reykjanes Ridge. Remember, the symmetry of the patterns on either side of the ridge was used as "proof" of seafloor spreading. Would this claim of near-perfect symmetry hold up when scrutinized by statistical analysis? The researchers used standard statistical formulas to calculate the coefficient of correlation for data sets taken on each side of the ridge. The main purpose of calculating the correlation coefficient is to avoid subjectivity in evaluation of data. The math either works or it doesn't. The calculated coefficient ranges from +1 to -1, with the zero point meaning "no correlation." and 1 indicating a high degree of correlation. Fifteen lines were drawn across the ridge (close to perpendicular) with an equal number of sample sites on each side. The number of samples for each line ranged from 80 to 150, so there was ample data to work with. The correlation coefficients for each line ranged from -.27 to .63, with overall average correlation of .17, with a standard of deviation of .23.[70] A coefficient this low gives us very little confidence that the scientists who first evaluated the ridge data were being entirely objective. They may have seen what they wanted to see.

Over the years, various other objections have been raised but it remains unclear how much weight these objections carry, as little follow-up was done after the initial publications. For example, in 1993 a mathematician claimed that the patterns of the variations in magnetic intensity are suspiciously similar to "band limited noise" observed in harmonic analysis. It looked to him like a system that is varying randomly but constrained to vary neither too slowly nor too quickly with distance. He used a computer to randomly generate a sequence of values with closer spacing than actual magnetic anomalies. The fine variation was then filtered by a model of the measurement process. When he compared his simulated anomaly profiles with examples of real profiles, he found that the general appearance of the simulated anomalies was remarkably similar to the actual profiles.[71]

The accuracy of radiometric dating has also been challenged. When samples of recently cooled lava from Mt. St. Helens (erupted in 1980) were tested using radiometric techniques (K-Ar), the minerals registered radiometric dates ranging from 340,000 years to 2.8 million years.[72] With results like this, how can we be sure of the accuracy of the ages assigned to the basalts on the "young" Mid-Atlantic Ridge?

It has also been suggested that the low points in the magnetic data (e.g. the lowest points on the graph on page 55) represent merely low intensities, not reversals of polarity. However, the process of data collection by a magnetometer does involve sensing the direction of the field. When an opposing field is detected, this is recorded as a subtraction from the overall field, thus becoming a "negative" anomaly on the graph. Unless hard evidence can be provided that pole reversal is impossible, we should assume that the magnetometer data is valid.

Other objections are related to the origin of the earth's magnetic field. Since we can't directly observe the inner earth, this topic will always remain theoretical. Because this is such an immense topic and requires expert knowledge of electricity and magnetism, it is beyond the scope of this book.

MORE QUESTIONS TO THINK ABOUT

Wouldn't magma at a spreading ridge follow the path of least resistance?

Actually, geologists have already thought about this, which is why they have backed away from "ridge push" as a primary mechanism of moving continental crust. Any magma welling up at the ridge would follow the path of least resistance, which in this case would be up into the water. Pushing apart two continental plates is a lot of hard work and rising magma is a very weak pushing force. Magma would never "choose" to do the hard work of pushing rock when it could simply escape upwards into the water. Magma will find cracks or weak spots in the rock and come out through them. The heat of the magma would probably melt the rock in the cracks, making them larger and allowing even more magma to escape.

How could magma at a spreading ridge possibly have enough hydraulic force to pull or push a giant land mass?

The force that hydraulic systems (e.g. pistons) are able to generate relies on the fact that the fluid is constrained inside a confined area. Magma under spreading centers, according to PT diagrams, is in an open system, not a closed system. Can PT show how their open hydraulic system is able to provide the unimaginable forces they claim?

How did these sedimentary rock layers bend?

Rocks don't bend smoothly under pressure; they crack and crumble. How can PT explain formations such as the one shown here? This formation is made of brittle sedimentary rocks. If the rocks had experienced extreme pressure due to plate motion, they should have turned into metamorphic rocks. (Could these layers have been semi-soft and pliable when they were compressed, then hardened afterwards?)

There are three places on the ocean floor where a trench meets a ridge: (50.5°N/130°W, 20.5°N/107°W, and 46.3°S/75.7°W). What is going on at these places?

The mantle material under these locations can't be going both up and down at the same time.

If the outer core contains much radioactivity (which is assumed to provide the source of earth's internal heat), and if molten material is rising from the outer core and coming out at seafloor spreading sites, then why is there almost no radioactivity in the oceanic crust?

Why has earthquake activity been observed to nearly double during low tides? [73]

If earthquakes are caused by plates rubbing against each other, then what is the connection to tides?

What causes large earthquakes to occur on the interior of a plate, far from any convergent or divergent border?
Very large quakes that occurred in the middle of a plate include Lisbon, Portugal (1755), New Madrid, Missouri (1811, 1812), Charleston, South Carolina (1886), and on the Antarctic plate (March, 1998). If earthquakes result from slippage between plates, what was the cause of huge intraplate quakes?

Some Benioff zones lie 80 to 150 km landward of the trench. What is happening under those 80 to 150 km

Why don't other planets have plate tectonics?
If all the planets were formed by (essentially) the same process, then why is the earth so different?

Reversals of plate motion have been recorded, especially during "slow slip" earthquakes.[74] How could a plate possibly reverse its motion if mantle convention and subduction are the driving forces?

Where are the missing ash layers from all the ancient volcanic activity?
Volcanoes usually produce a considerable amount of ash. We do find ash layers in the geologic record, but they tend to be in upper layers. If volcanic activity was a primary characteristic of the early earth, why do we not find many ash layers in pre-Cambrian or Archaeozoic rocks?

Why are there almost no meteorites or craters in radiometrically old rock layers?
Meteors and craters are found almost exclusively in the top layers—the layers that are said to be young.

Why have we found so many pieces of continental crust in various parts of the ocean?
The largest piece found to date is Zealandia (east of Australia), which is so large that it has been classified as a new continent.

If the Cascadian margin (the coastline along Vancouver, B.C. and Washington state) is classified as a convergent boundary, why is there no obvious trench and very few earthquakes in this region?
Bathymetry expert Christian Smoot has used a number of deep sea scanning devices to search for evidence of a trench along the Cascadian margin. Despite the claim that one exists, he is adamant that neither his sonar images nor his seismic data show a trench in this location.[75]

For several centuries, people have written about strange electrical activity that occurred right before an earthquake. Now we know that pre-quake electrical activity occurs even in the ionosphere. Does PT have an explanation?

IF PLATE TECTONIC THEORY IS SO FAULTY, HOW DID IT BECOME SO POPULAR? WHAT HAPPENED?

The information in this chapter was gleaned primarily from an essay entitled "Geoscientific Urban Legends" by Norwegian geologist Karsten Storetvedt, published in the New Concepts in Global Tectonics Journal, volume 3, number 4, December 2015. Some readers will want more details than are provided here. You can find the full essay at www.ncgtjournal.org.

Karsten Storetvedt was a college student studying geology at the University of Bergen (Norway) from 1957 to 1962. He stayed in academia afterwards and began teaching college geology classes. Because of his university position, he was right there in the midst of the tectonic revolution taking place in the early 1960s. He met some of the original PT theorists, and worked with people who knew them. He was an eyewitness to the events that led up to the official acceptance of PT theory in 1966. At first, he was heavily influenced by the new ideas about global continental drift, and brought each new idea to the students in his classroom, However, as he began to study the details for himself, especially details about paleomagnetism, he saw the theory running into quite a few technical problems. If he tried to discuss these problems with other professional geologists, they quickly shut him down. His naive assumptions about scientists being objective seekers of truth soon vanished. Disillusioned, and ostracized by his academic peers, Storetvedt eventually resigned his post from the university and became an independent researcher. In 2015, he decided to write about the people and events surrounding the formulation of PT theory in the mid-1960s, and provide us with answers to the questions: "What happened? When and how did things get so off-track?"

The first edition of Alfred Wegener's book, *The Origin of Continents and Oceans,* was published (in German) in 1915. Subsequent editions added more details to the theory and the first English edition was printed in 1924. The last edition of the book was printed in 1929 and was not translated into English until 1962. We assume that all the original theorists read this book since their work built upon it, but apparently this was not the case. Storetvedt interviewed some of the original theorists during their retirement years, and they were willing to talk candidly about the people and events of the mid-60s. Storetvedt was shocked to learn that most of the theorists had never actually read Wegener's book. If they had, they would have been introduced to problems that Wegener saw with his own theory. Instead, the theorists relied on secondary sources, especially these two books: *Principles of Physical Geology* by Arthur Holmes, and *Our Wandering Continents* by Alexander du Toit. These authors were supporters of Wegener during the years when his ideas were sharply criticized, or even ridiculed. When Holmes and du Toit wrote their own books, they included some of Wegener's most important ideas, but they didn't bother telling their readers the problems that Wegener himself had uncovered. For example, paleoclimate evidence from Antarctica did not support the theory, nor did the geology of the Azores archipelago. Holmes and du Toit presented a very sympathetic, "polished" version of the theory, without any discussion of critical counter evidence.

The dependency went both ways—in his later editions, Wegener made use of research done by du Toit during the 1920s. Du Toit studied the rocks on the east coast of South America and the west coast of Africa and said that there were enough similarities that he felt confident in asserting that these continents had been much closer in the past. However, he also said that to account for some notable differences in the coastal rocks, the pre-drift separation distance had to be 400-800 km. Du Toit never proposed that these landmasses were joined together as part of a super continent. Wegener took liberties with du Toit's idea, however, and made the claim in his own book that the two continents fit together like the edges of a torn piece of newspaper.

Wegener, in his 1929 edition, did discuss the problem that the Azores posed to his theory. He admitted that geological evidence pointed to these islands having a continental substratum. Continental-type rocks such as granite, mica-schist, quartzite, sandstone, and limestone had been collected from the large plateaus around these islands for decades. Wegener knew that if there was a sizable piece of continental crust in the middle of the Atlantic (the area all around the islands) it would be much more difficult to explain how the Atlantic Ocean opened up. Another problem that Wegener seemed to be aware of, but disregarded, was paleoclimate evidence that Antarctica had once been a tropical or sub-tropical region. His assembly of the super continent Gondwana had all the continents grouped around Antarctica in its polar location.

Despite the fact that Wegener himself discussed counter evidence that could be used to argue against his theory, geologists of the 1950s and '60s were much less critical. Storetvedt remembers that during this time discussing these counter arguments was seen as "old fashioned" and a waste of time.

Until the 1950s, the general consensus among geologists was that the crust of the earth was contracting or shriveling like the skin of a drying apple, and this was causing mountains and trenches to form. They scorned the idea of lateral motion of the continents. Even Tuzo Wilson, who would become one of the foremost promoters of plate tectonics, was on record in 1959 as a supporter of the contraction hypothesis. The shift came about because of a related field: paleomagnetics.

During the 1950s, there was a lot of research being done in a new field of study called *paleomagnetism*. Someone discovered that not only was there variation in the magnetism of rocks in different areas of the world, but there were also variations as you dug straight down into the rock. It seemed that the layers of rock had recorded where the north pole was in the past. They took samples from layers that had been assigned to various radiometric ages, and from them figured out where the north pole would have been during that radiometric era. When all the points were put onto a globe, you could connect the dots to make a curve, showing how the north pole had apparently "wandered" over time.

This concept of "polar wander" soon became a very hot topic. At first, the data was taken at face value and they sought an explanation for how the magnetic pole could change over time. Then more data came in. The same experiments were performed on all the continents. The polar wander curves for Europe, North America, and Asia were completely different. The north pole could not have been in more than one place at one time, so a new interpretation of the data was needed. Perhaps the continents had moved, not the poles? Didn't someone have a theory about that? Yes, there it was—a theory about continental drift, already neatly packaged into a published book. Wegener's theory was suddenly at center stage. Storetvedt recalls that they never put any effort into thinking of alternate explanations. He would have liked them to consider a slight in situ rotation of the continents, as this would have been able to account for the variation in polar wander paths. But once they latched onto Wegener's ideas, they never let go.

Other factors then came into play. For example, there were two leading, and competing, research teams in Britain: one led by Patrick Blackett and the other by Keith Runcorn. At first, the Runcorn group was supportive of polar wander as an actual phenomenon. Runcorn wrote that true polar wander was the simpler hypothesis (compared to continental drift) and that it was easier to make this hypothesis fit with the [then] current understanding of geodynamics. However, the dramatically different polar wander curves were luring most geologists, including Blackett, in the direction of continental drift theory. India's polar wander curve indicated that it had been much further south in the past. The northward movement of India was a key part of Wegener's hypothesized break-up of a southern supercontinent.

In 1992, Storetvedt was able to interview Ernie Deutsch, one of the principle investigators of the India magnetism project and an ally of the Blackett group. Storetvedt asked him what was going through their minds as they decided to adopt the continental drift theory. Ernie admitted that in retrospect their conclusion had been affected both by ignorance of geological data and by rivalry with the Runcorn group. None of the researchers that Ernie worked with had actually read Wegener's and du Toit's books; they just assumed that someone else had already vetted them. Ernie also confessed that the science of magnetism was new and obscure, and being associated with an exciting, new idea (continental drift) would give paleomagnetism a boost and help create funding for its researchers.

Not long after the Blackett group endorsed continental drift, Keith Runcorn did an "about face" and not only endorsed continental drift but also became its most ardent promoter. Runcorn went around the world on lecture tours, attempting to break down the opposition to "new global tectonics." Obviously, he eventually won the day, and was able to steer the worldwide geological community towards complete acceptance of continental drift. The powerful inertia of the plate tectonic movement had begun.

In the early 1960s, continental drift was much more popular in Britain than in America. A few "pioneers" (Hess, Heezen, Dietz) were already on board, but the theory had not gained general acceptance. A key event that helped to tip the scale in favor of drift was a symposium in 1964 sponsored by the Royal Society and organized by Blackett, Runcorn, and Edward Bullard. At this meeting, Bullard introduced his now-famous "fit" of the continents. Despite the fact that Bullard edited out all of Central America, he was able to impress his audience by claiming that his work was based on objective mathematical methods such as Euler's theorem for motions on a sphere, and on work he had done using a computer, which was cutting edge technology at that time. Despite the fact that it was obviously an abuse of mathematics, the "Bullard's fit" became one of the most cited works in geoscientific literature.

One of the first American converts was Tuzo Wilson. Like his British counterparts, Wilson's conversion to drift theory was almost overnight. In 1959, Wilson wrote a paper in which he rejected both continental drift and mantle convection. A year later he was lecturing about mobile continents. By 1963 he had endorsed not only Wegener's continental drift, but also mantle convection and seafloor spreading. In his 1959 paper, he had confidently concluded that the mid-ocean ridges were very old and had never moved. Now, in 1963, he was claiming that the ridges were very young and that the seafloor was in constant motion. He proposed that young, active volcanoes would be found near the mid-ocean ridges, and older, less-active (or extinct) volcanoes would be found further from the ridges. This was already known to be false in 1963, and additional data since then has confirmed that ridges are the least likely place in the ocean to have young volcanoes (with the exception of Iceland, an anomaly if there ever was one). Despite the fact that Wilson's claims about the ridges could easily be proved wrong, his paper became one of the most cited works in the following decade.

Plenty of other American geologists were happy to add fuel to the plate tectonic fire. In 1960, Bruce Heezen wrote a paper suggesting that mid-ocean ridges were places where new seafloor was forming. Harry Hess and Robert Dietz soon joined the party and began embellishing the story about seafloor spreading. However, behind closed doors, some of them had doubts about the storyline they were creating. Dr. Brown, the author of one of the alternative theories in the next section, was a young colleague of Bob Dietz in the 1970s, and spent an afternoon with him once a week, for about a year. They would discuss a wide range of geological and geophysical topics. When Brown would ask Dietz to explain the mechanisms of seafloor spreading or to propose possible solutions to some of the obvious problems with PT theory, Dietz would candidly admit, "To be honest, I don't have a clue." After hearing this admission on a regular basis, it's no surprise that Brown was emboldened to offer an alternative theory that could explain the geological features that stumped the PT theorists.

SOME ALTERNATIVE THEORIES

In 1897, Dr. Thomas Chrowder Chamberlain, former president of the University of Wisconsin and the first head of the Geology Department at the University of Chicago, published a famous paper in the Journal of Geology in which he warned researchers not to let one hypothesis dominate their thinking. Instead, he said, they should always be seeking multiple hypotheses. By testing competing hypotheses we sharpen our analytical skills, develop thoroughness, reduce bias, and learn to think independently. He thought the danger of teaching only one explanation was especially great in the earth sciences, where we study so many things that cannot be directly observed. Unfortunately, the world of geology has done exactly what Dr. Chamberlain advised against, letting one hypothesis become so dominant that it is taught as fact, and students are kept from exploring any other options. There's no reason that several theories can't be presented to students, each with its supporting evidence and its problems, and allow the students to use their critical thinking skills to decide for themselves which they think is the most likely scenario.

Here are short summaries of some alternatives to Plate Tectonics.

Global Wrench Tectonics

This idea comes from Norwegian geologist, Karsten Storetvedt. Looking back on his formative college years in the 1960s, he says this about himself: "I believed scientific theories were products of meticulous observation and careful experiment—building up objectively proven knowledge, step by step. The fact that science is also a human activity with all its "binding factors"– including the complex web of social, cultural and psychological circumstances—was not in my mind. With the benefit of hindsight I gradually came to realize that I had grown up with the erroneous notion that the scientific establishment...welcomes progress, which in fact is the opposite of what is generally true." While researching paleomagnetism in the late 1960s, Storetvedt began finding out how much the data had been stretched and tweaked in order to fit the PT model. He was shocked to find out that many scientists would rather adjust their data than their hypothesis. By 1969 he had concluded that there was something fundamentally wrong with plate tectonics.

While working on the Great Glen Fault in northern Scotland, fall 1972, he became aware of "wrench deformation." The observed discrepancies between the North Atlantic polar wander paths could have arisen from in situ rotations of Europe and North America. It seemed that lateral motions of continental drift were not required to account for the global pattern of paleomagnetic directions. During this time he also began teaching classes, and he introduced his students to different ways of interpreting data. The students responded with enthusiasm, but members of the teaching community accused him of being disloyal and without a will to adopt a co-operative attitude.

His academic career became increasingly difficult as the professors and geologists he interacted with became increasingly hostile. In 1989 he went on sabbatical in England, and found himself in a relaxed environment where he was able to spend a lot of time thinking. His epiphany came at this time: Global Wrench Tectonics.

He realized that it would be virtually impossible to combine revolutionary big-scale thinking with traditional academic progress/activities, so he decided to become an independent researcher and visiting scholar. He has traveled all over Europe and North America giving talks about his new tectonic theory.

His theory proposes that the Earth started out with a global crust made of granitic-type rock which gradually thinned over time. The interior, under the crust, had lots of water locked into the crystal structure of the minerals. Some of this water came out (in a process he calls degassing) and became "supercritical" in the high-pressure environment. Supercritical water pressing upward provided the force for uplift of mountains. Pressure from below also caused the crust under the oceans to become thinner than continental crust. Vertical mass transfer (formation of mountains, for example) has affected planetary rotation, giving rise to episodic changes of spin rate and of spatial reorientations of the Earth, providing a mechanism to explain polar wander. These rotational "jerks" are the triggers for tectonic processes. He thinks the evidence (especially the deep roots of the continental interiors) favors stationary continents.

Crustal mobility, which is suggested by paleomagnetic and space geodetic measurements, is primarily limited to in situ changes of the continental blocks. At times of changes in rotation, the entire lithosphere has been subjected to wrench deformation (torsion). Sudden degassing and the resulting changes in mantle pressure created faults and fracture zones and resulted in sudden subsidence of various parts of the globe. Volcanism is the result of shifts in the lithosphere, and do not involve any deep processes in the mantle. The similarity between the coastlines of the west and east Atlantic Ocean is due to parallel fractures zones along the subsiding ocean crust.

If you want to know more about this theory, go to www.storetvedt.com.

Surge Tectonics

This theory was first proposed by Howard and Aurthur Meyerhoff in 1977, though the theory has been expanded and adapted since then. Surge tectonics is based on the concept that the lithosphere contains a global network of magma channels called surge channels. A surge channel is a conduit for magma to rise from the asthenosphere (the partially molten layer under the lithosphere). The idea is that the lithosphere and the asthenosphere form a giant hydraulic system, with a containment vessel (the surge channel system), a fluid (magma), and a trigger mechanism (collapse of the lithosphere into the asthenosphere when the latter becomes too weak to support it). The hydraulic press system is driven by the cooling and shrinking of the lithosphere over time, and is linked to gravity and to the rotation of the earth. There is a differential "lag" that occurs between the top of the lithosphere and the bottom of the asthenosphere, as a result of the earth's rotation. This lag produces a relative eastward movement of magma. Another contributing factor is that magma rising from the asthenosphere expands its volume, reducing gravitational attraction in that part of the surge channel.

The mid-ocean ridges are the active trunks of the surge channels. They feed the other channels, such as the ones that have produced the linear seamount chains. The empty, collapsing, inactive surge channels are the fracture zones. Occasional dribbles of lava in an inactive channel will produce a later seamount. Fracture ridges, fracture valleys, and seamount chains are seen as surface evidence created by the same underlying cause (surge channels) so this is why we have megatrends. Where the channels intersect, they create vortex structures. Positive vortex structures are regions of upwelling, and the features residing there are called rises and plateaus. Various groups of islands are thought to be the result of vortices.

The entire hydraulic press phenomenon is linked to gravity. Magma rising from the asthenosphere undergoes inversion to lighter, less compact phases. This expansion decreases density, which reduces the gravitational attraction within the region of the surge channel, and this affects barometric conditions. Surge tectonics suggests that this is what happens during an El Niño. Magma rises from the asthenosphere to the East Pacific Rise through the lithosphere. The slight weakening of the gravity field is translated to the atmosphere as the pressure drops. Surge tectonics advocates believe that the goings-on in the surge channels are what drives global warming or cooling trends, as well as being the cause of earthquakes and volcanoes.

Surge Tectonics has advocates in every major country. They maintain an online journal where monthly reports and papers are published. You can learn more by going to www.ncgtjournal.com.

Shock Dynamics Geology

This idea was conceived by John Michael Fischer, who posts details of his theory on his website: newgeology.us. His theory proposes that the earth was struck by a giant asteroid, creating a globe-shattering impact. The center of the impact site, he says, was in the channel between Africa and Madagascar, though at the time of the impact, neither Africa nor Madagascar looked like they do today. Like plate tectonic theory, he believes that there used to be one large super-continent that was split into pieces. However, unlike PT theory, Fischer proposes that the split of the continents and their subsequent movement was caused by a huge shock wave (created by the giant impact).

Describing exactly what the shock wave did is a bit of a challenge. Fischer has animations on his site that he uses to supplement his text. When the impactor hit, it punched a hole right where the Mozambique Channel is. The shock energy pushed Africa westward as well as crumpling its eastern side, creating Africa's mountains and volcanoes. The shock wave continued through the globe, with the vibrations causing interesting, but temporary effects, (becoming a Bingham fluid), allowing the rock slabs to move more easily than they would have under normal conditions. A crack opened up between Africa and what would eventually become South America. This crack today has become the mid-Atlantic ridge. The crack went north and split the future North America from future Europe. The Americas went sliding off to the west, with the space between (now ocean) experiencing massive lava flooding, and leaving an active ridge in the middle. When North and South America skidded to a stop, their western edges crumpled and the Rockies and Andes mountains formed. Central America "unfolded" from between North and South America during the continental sliding.

When India slammed into Asia, it did so at an angle that caused Asia to move slightly counterclockwise. Most of the energy was absorbed in the formation of the Himalayan Mountains, but there was enough energy left to push Asia north and east. This counterclockwise rotation of eastern Asia was felt by eastern Europe and the Ural Mountains. (The pivot point for all this rotation was where Iran is today.) As Europe felt the counterclockwise push, this created its southern mountain ranges, such as the Alps. At the same time all this was happening, Australia was sent directly east of the impact site.

The Appalachian Mountains and the mountains of eastern Brazil are explained as "impulse" mountains, formed when the shock impulse first arrived at that area. Collision mountains were formed when two landmasses collided, such as Alaska into North America, and India into Asia. (Alaska would have originally been next to what became Japan and Kamchatka.) The "roots" of mountains would have not originally been present in the pre-impact earth, but were created as the mountains began to rise.

The trenches of the Pacific, he says, are the end products of the shock wave that went eastward from the impact site. The wave traveled almost in fluid form until it ran out of energy and solidified, seen now as the long north-south trenches of the Pacific. Small pieces of continental crust may have broken off and scattered, leaving a curved "wake" behind them, which we see now as curved trenches. Alaska would have created the Aleutian trench as it was flung from northern Asia. Geologically, Alaska is a very diverse and complex area and is geologically dissimilar to North America.

There were already several smaller impact craters on the super-continent before this huge impactor hit. Hudson Bay is given as an example, though the bay was greatly enlarged as North America slid after the impact. The East Pacific Rise is another feature believed to have been created before the great impact.

Shock dynamics attributes the slow plate motion we see today to the gravitational pulling forces of the sun, moon and planets. The plates shift around slightly due to this pulling effect.

You can find out more by going to http://newgeology.us, or by going to You-Tube and searching for "Shock Dynamic Geology."

Hydroplate Theory

The Hydroplate Theory was first published by Dr. Walter Brown in the 1980s in a book titled *In the Beginning*. It took over two decades for Dr. Brown to realize all the ramifications of the global cataclysms his theory proposes. As he added more details to his theory he would publish a new edition. His book is now in its 9th edition and covers not only global tectonics, but also the origin of comets and meteors, the origin of radio-activity, and a detailed explanation of how the Grand Canyon formed.

The Hydroplate Theory makes a few simple starting assumptions.

1) CRUST: The crust of the earth was a solid shell made of granitic rock, at least 15 km (10 miles) thick but possibly up to 95 km (60 miles) thick. The crust held many bodies of water, but nothing as large as the oceans we have today. At various points, the crust sagged down, creating "pillars" that touched the mantle.

2) WATER: Underneath the crust was a layer of water several kilometers deep. The water was rich in minerals and dissolved carbon dioxide. It was under extreme pressure.

3) MANTLE: The mantle was much like it is today: solid, made of olivine and peridotite.

The water under the crust experienced "tides" due to the moon's gravity, and this created "tidal pumping" which heated the water until it became "supercritical"—a phase of matter that contains an immense amount of energy.

A crack developed in the granitic crust; it spread from both ends and quickly encircled the globe. Supercritical water jetted out of the crack, pulverizing the granite walls on either side of the crack. Water and chunks of rock escaped earth's gravity and were hurled into space. All the ejected water and debris eventually coalesced (due to their own gravity) as they traveled away from the earth. They became comets and asteroids, many of which settled into orbits around the sun. This explains why comets and asteroids have a chemical and mineral composition similar to the earth's. Dr. Brown predicted that at least some asteroids would be covered with rounded boulders. (This prediction was confirmed in 2014 on comet 67P.)

Enough water came up from below the crust that eventually the entire surface of the globe was covered. The water was saturated with minerals, creating perfect conditions for fossilization. This event explains the mass extinctions we see in the fossil record, including whale fossils in areas that are now deserts. It also explains how even delicate, soft-bodied animals could be preserved; they were quickly and gently buried in mineral-saturated sediments, many of which turned into rocks such as limestone.

The "limestone problem" is easily solved by HPT. Chemical production of limestone ($CaCO_3$) produces CO_2. If the amount of limestone we see today had all come from evaporating oceans, it would have produced enough CO_2 to kill all life on the planet hundreds of times over. However, if the $CaCO_3$ formed under the crust and simply came to the surface, there would be no massive release of CO_2. Dr. Brown suggests that the hot, salty water found at the bottom of the Kola and KTB boreholes came from this "primordial" salty water under the crust.

The crack eventually widened so much—possibly several hundred kilometers— that the mantle underneath began to bulge upward. (This is called isostatic rebound and is a phenomenon often observed in the floor of quarries after much rock is removed.) As the mantle bulged, tension cracks began to form. The first tension cracks were along the length of the crustal crack. These became the fracture zones along the mid-ocean ridges. Then, as the split widened, the mantle began bulging in that direction, and tension cracks popped open between the fracture zone cracks, creating the central ridges. The movement of the mantle rock caused a lot of frictional heating and magma came pouring out in many places, creating the flood basalts that pave much of the ocean floors.

Even though water was escaping from under the crust, there was still enough water to provide an almost friction-free ride for the continents. As the mantle bulged up, the continental plates (the "hydroplates") slid away from the ridge. The water cushion didn't last long and the bottoms of the plates soon hit the mantle below and came to a screeching halt. This sudden deceleration and massive friction caused the leading edge

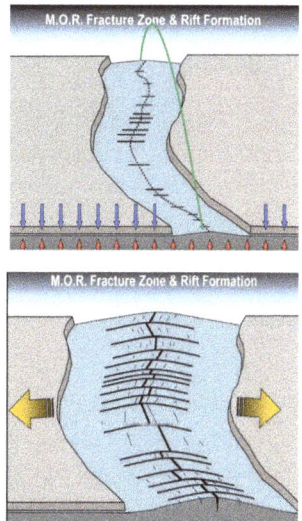

M.O.R. Fracture Zone & Rift Formation

M.O.R. Fracture Zone & Rift Formation

of the American plate to crumple, creating the Rocky Mountains and the Andes Mountains. The lagging edge also experienced some friction, and small mountain ranges (e.g. the Appalachians) were formed. In the other hemisphere, the land masses moved away from both the Atlantic ridge and the Indian Ocean ridge, causing Antarctica to move south. The sudden deceleration of the northward-bound Asian plate produced the Himalayas. Over the following centuries, the thickened crust under the mountains sank into the mantle creating the mountains' roots. Because the water is no longer under the crust, we are left with an apparent mystery—we see large plates that seem to have moved, but they are now resting on rock that would seem to prevent any large scale movement.

Since "nature abhors a vacuum," the bulging of the mantle in the Atlantic caused all the rock underneath it to shift in that direction. Dr. Brown calculated that this amount of shifting deep in the earth could easily melt the rock of the inner earth and create what we now call the inner and outer core. On the Pacific side of the world, this shifting was felt as a downward pull on the crust. The crust buckled and cracked, creating the arc and cusp shapes of the Ring of Fire. Deep fractures along the rim created the Benioff zones. The Pacific crust then sank down onto the mantle, creating huge faults and fractures throughout the mantle. Melted mantle rock poured out as flood basalts, paving the floor of what is now the Pacific Ocean. When chunks of the granitic crust melted, their magma produced the steep, cone-shaped seamounts. (Only magma from granitic rock is thick enough to make steep cones.) Long fractures made lines of seamounts. In a few places, the mantle buckled deeply, and trenches were formed. It's not a coincidence that the deepest trenches of the world are almost directly opposite the Mid-Atlantic Ridge.

Dr. Brown makes a distinction between drifting and shifting. The continents drifted quickly as a one-time event. Now they are simply shifting from time to time, in small jerking motions, as the spinning globe tries to even out its mass and become perfectly round again. The shifts are almost always in the direction of the globe's mass deficiencies—the trench zones, usually the ones in the western Pacific. Shifts occur along fault lines and often result in earthquakes and volcanic activity. The root cause of all the shifting is the slipping of fractured rocks deep in the mantle. Slippage at shallow depths creates magma that rises to the surface as volcanoes. Slippage at great depths creates magma that drains into the outer core. As the outer core enlarges, it puts pressure on the mantle above it and this creates even more slippage. Other theories assume that shifting plates produce earthquakes. Hydroplate Theory says that earthquakes make the plates shift.

Radioactivity was created in areas of the granitic crust that were most affected by the sudden deceleration. The piezo-electric properties of quartz combined with a high-voltage process known as Z-pinch, as atoms were torn apart and then recombined into super-heavy elements, many of which fell apart quickly, but some of which turned into uranium and thorium. This might help to explain the mystery of why radioactivity is found primarily in continental crust and preferentially in granite.

Dr. Brown's book can be read online at www.hydroplate.org.

END NOTES

[1] Lyustikh, E. N. 1967. "Criticism of Hypotheses of Convection and Continental Drift." *Geophysics. Journal of the Royal Astronomical Society* 14. pp. 347-352. (https://academic.oup.com/gji/article/14/1-4/347/742279)

[2] Beloussov, V. V. 1970. "Against the Hypothesis of Ocean-floor Spreading." *Tectonophysics* 9. pp. 489-511. Amsterdam: Elseveir Publishing Company.

[3] Meyerhoff, A. A., Howard A. Meyerhoff, and R. S. Briggs. 1972. "Continental Drift, V: Proposed Hypothesis of Earth Tectonics." *The Journal of Geology* 80, no. 6: pp. 663–92. (http://www.jstor.org/stable/30080140)

[4] Smoot, N. Christian. 2004. *Tectonic Globaloney*. Xlibris Corp. pp. 160-162.

[5] Stern, R. J. 2002. "Subduction zones". *Rev. Geophys.* 40. doi: 10.1029/2001RG000108.

[6] Smoot, 2004. p. 34.

[7] Graphics shown are by Bryan Nickel, and are adaptations of Dr. Walt Brown's work on page 422 of *In the Beginning; Compelling Evidence for Creation and the Flood, 8th Edition*. 2008. Center for Scientific Creation. Phoenix, Arizona. Pages 422, 423.

[8] Gudmundsson, August. 2013. "Great Challenges in Structural Geology and Tectonics." *Frontiers in Earth Science.* 12 November 2013 (https://doi.org/10.3389/feart.2013.00002)

[9] Pratt, David. 2000. "Plate Tectonics: A Paradigm Under Threat." *Journal of Scientific Exploration*, Vol. 14, No. 3. p. 308.

[10] Menard, H. W. 1964. "Darwin Reprise." Journal of Geophysics. Res. 89. pp. 9960-9968

[11] Smoot, N. Christian and Robert E. King. 1997. "The Darwin Rise Demise: the Western Pacific Guyot Heights Trace the Trans-Pacific Mendocino Fracture Zone." *Geomorphology* 18, 223-235.

[12] Dobson, M. R. and Dennis O'Leary. 2019. "Sediment delivery and depositional processes along the eastern Aleutian Trench." *Journal of the Geological Society,* Vol. 176, pp. 1076-1092.

[13] Ibid.

[14] Angiboust, Samuel, Arnel Menant, Taras Gerya, and Onno Onchen. "The Rise and Demise of Deep Accretionary Wedges: A Long-term Field and Numerical Modeling Perspective." *Geosphere* 18, number 1. Published by the Geological Society of America.

[15] Beloussov, 1970.

[16] Tarbuck, Edward J. and Frederick K. Lutgens. 2017. *Earth: An Introduction to Physical Geology 12th edition.* Pearson Education, Inc.pp. 170-173. (A popular college text)

[17] Marshak, Stephen. 2016. *Essentials of Geology, 5th edition.* W. W. Norton & Co. New York, London.pp. 132-133. (A popular college text)

[18] Brown, Walter. 2022. *In the Beginning; Compelling Evidence for Creation and the Flood 9th edition.* Phoeniz, Arizona: Center for Scientific Creation.. pp.175 (Accessed using this web address: https://hpt.rsr.org/flipbook/)

[19] Brudzinski, Michael R., Clifford H. Thurber, Bradley R. Hacker, and E. Robert Engdahl. 2007. "Global Prevalence of Doube Benioff Zones." *Science,* Vol. 316.

[20] van Hunen, Jeroen; van den Berg, Arie P; Vlaar, Nico J. 2004. "Various mechanisms to induce present-day shallow flat subduction and implications for the younger Earth: a numerical parameter study". *Physics of the Earth and Planetary Interiors. Plumes and Superplumes.* 146 (1–2): 179–194. doi:10.1016/j.pepi.2003.07.027.

[21] Skinner, Steven M. and Robert W. Clayton. 2013 (June 1). "The lack of correlation between flat slabs and bathymetric impactors in South America." *Earth and Planetary Science Letters.* 371–372: 1–5. (doi:10.1016/j.epsl.2013.04.013)

[22] van Hunen et al., 2004.

[23] Manea, Vlad C., Marta Pérez-Gussinyé, and Marina Manea. 2012 (Jan. 1). "Chilean flat slab subduction controlled by overriding plate thickness and trench rollback". *Geology.* 40 (1): 35–38. doi:10.1130/G32543.1. ISSN 0091-7613.

[24] Knapmeyer, Martin. 1999. "Geometry of the Aegean Benioff Zones." *Annal di Geofisica*, Vol. 42. N. 1, Feb 1999.

[25] Beloussov, 1970.

[26] Jackson, James. 2005. "Mountain roots and the survival of cratons." *Astronomy & Geophysics*, Volume 46, Issue 2, April 2005, Pages 2.33–2.36, https://doi.org/10.1111/j.1468-4004.2005.46233.

[27] Meyerhoff, Arthur, et al. 1996. *Surge Tectonics —A New Hypothesis in Global Geodynamics.* Dordrecht: Kluwer Academic Publishers. page 1

[28] Pavlenkova, N.I. 1996. "General features of the uppermost mantle stratification from long-range seismic profiles." *Tectonophysics, v.* 264, p. 261-278.

[29] Lerner-Lam, A.L. 1988. "Seismological studies of the lithosphere." *Lamont-Doherty Geological Observatory Yearbook,* p. 50-55.

[30] Aumento, F., and B. D. Loncarevic. 1969. "The Mid-Atlantic Ridge near 45°N., III. Bald Mountain." *Canadian Journal of Earth Sciences,* 6, 11-23.

[31] Wanless, R. K., R. D. Stevens, G. R. Lachance, and C. M. Edmonds. 1968. "Age determinations and geological studies. K-Ar isotopic ages, report 8." *Geological Survey of Canada,* Paper 67-2, pt. A, pp. 140-141.

[32] Meyerhoff, A. A., Taner, I., Morris, A. E. L., Agocs, W. B., Kaymen-Kaye, M., Bhat, M. I., Smoot, N. C., & Choi, D. R. 1996. *Surge Tectonics: a New Hypothesis of Global Geodynamics.* (D. Meyerhoff Hull, Ed.). Dordrecht: Kluwer.

[33] MacDougall, D. 1971. "Deep sea drilling: Age and composition of an Atlantic basaltic intrusion. *Science*, 171. pp. 1,244-1,245.

[34] Anderson, R. N., Honnorez, J., Becker, K., Adamson, A. C., Alt, J. C., Emmermann, R., Kempton, P. D., Kinoshita, H., Laverne, C., Mottl, M. J., & Newmark, R. L. 1982. "DSDP hole 504B, the first reference section over 1 km through Layer 2 of the oceanic crust." *Nature*, 300, 589-594.

[35] Smoot, 2004, pg. 41.

[36] Melson, W. G., Hart, S. R., & Thompson, G. 1972. "St. Paul's Rocks, equatorial Atlantic: petrogenesis, radiometric ages, and implications on sea-floor spreading." In R. Shagam, R. B. Hargraves, W. J. Morgan, F. B. Van Houten, C. A. Burk, H. D. Holland, & L. C. Hollister (eds.), *Studies in Earth and Space Sciences* (Memoir 132, pp. 241-272). Boulder, CO: Geological Society of America.

[37] Smoot, 2004. pg. 43.

[38] Smoot, N. C., & Meyerhoff, A. A. (1995). "Tectonic fabric of the Atlantic Ocean floor: speculation vs. reality." *Journal of Petroleum Geology*, 18, 207-222. (From a quote about a personal correspondence with Barrie Richards from Cambridge Univ. 1999.)

[39] Reynolds, P. H., & Clay, W. 1977. "Leg 37 basalts and gabbro: K-Ar and 40Ar-39Ar dating." In F. Aumento, W. G. Melson, et al., *Initial reports of the Deep Sea Drilling Project* (vol. 37, pp. 629-630). Washington, D. C.: U. S. Government Printing Office.

[40] Brown, 2022. p. 120.

[41] Lyustikh, 1967.

[42] Storetvedt, Karsten. 1997. *Our Evolving Planet: Earth History in New Perspective.* Bergen, Norway: Alma Mater.

[43] Smoot, N. C. 2015. *Marine Geomorphology*, 3rd edition. Portsmouth, New Hampshire: Mindstir Media. p. 65.

[44] Masalu, Desiderius. 2002. "Absolute migration and the evolution of the Rodriguez Triple Junction since 75 Ma." *Tanzania Journal of Science*. 28. 10.4314/tjs.v28i2.18358.

[45] Ibid.

[46] Urakawa, Satoru, Tatsuya Sakamaki, and Eiji Ohtani. 2006. "Anomalous compression of basaltic magma: Implications to pressure-induced structural change in silicate melt." *Research Frontiers.* pp. 113-114. www.spring8.or.jp/pdf/en/res_fro/06/113-114.pdf

[47] Rigden, S. M., T. J. Ahrens, and E. M. Stolper. 1984. "Densities of liquid silicates at high pressures." *Science* 226. pp. 1071-1074.

[48] Ohtani, E. 1983. "Melting temperature distribution and fractionation in the lower mantle." *Phys. Earth Planet.* Int. 33, pp 12-25.

[49] Agee, Carl B. 1998. "Crystal-liquid density inversions in terrestrial and lunar magmas." *Physics of the Earth and Planetary Interiors*, 107. pp. 63-74

[50] Christensen, Ulrich. 1998. "Fixed Hotspots Gone with the Wind," *Nature*, Vol. 391, 19 February 1998, p. 740.

[51] Marsh, Bruce D. 1980. "Island-Arc Volcanism," *Earth's History, Structure and Materials*, editor Brian J. Skinner (Los Altos, California: William Kaufman, Inc., 1980), p. 108.

[52] Gordon A. Macdonald et al., Volcanoes in the Sea, 2nd edition (Honolulu: University of Hawaii Press, 1983), p. 337.

[53] Hamilton, Warren. 2002. "The Closed Upper-Mantle Circulation of Plate Tectonics, in Plate Boundary Zones." *American Geophysical Union Geodynamics Series* Volume 30, eds. Seth Stein and Jeffrey T. Freymueller, pp. 359-410.

[54] Sheth, H. C. 1999. "Flood basalts and large igneous provinces from deep mantle plumes: fact, fiction, and fallacy." *Tectonophysics*. Volume 311, Issues 1–4, 30 September 1999, Pages 1-29.

[55] Foulger, G. et al., 2003; The Geological Society of London Great Plumes Online Debate.

[56] https://en.wikipedia.org/wiki/Kola_Superdeep_Borehole

[57] MacDonald, G. J. 1988. "Major Questions About Deep Continental Structures". In A. Bodén and K. G. Eriksson (ed.). *Deep drilling in crystalline bedrock*, v. 1. Berlin: Springer-Verlag. pp. 28–48. ISBN 978-3-540-18995-4.

[58] Emmermann, Rolf, and Jorn Lauterjung. 1997. "The German Continental Deep Drilling Program KTB: Overview and major results." *Journal of Geophysical Research*, Vol. 102, No. B8, pages 18,179- 18,201.

[59] Ibid.

[60] Clauser, Christoph, et al. 1997. "The thermal regime of the crystalline continental crust: Implications from the KTB." *Journal of Geophysical Research*, Vol 102. No. B8. August 10, 1997. pp. 18,417-18,441.
Additional confirmation of the chemistry of the water at the bottom of the bore holes: Prokofiev, V.Y., Banks, D.A., Lobanov, K.V. et al. "Exceptional Concentrations of Gold Nanoparticles in 1,7 Ga Fluid Inclusions From the Kola Superdeep Borehole, North west Russia." *Sci Rep 10*, 1108 (2020). https://doi.org/10.1038/s41598-020-58020-8

[61] https://en.wikipedia.org/wiki/Xenolith

[62] Luyendyk, B. P., and W. G. Melson. 1967. "Magnetic properties and petrology of rocks near the crest of the Mid-Atlantic Ridge." *Nature*, vol. 215, no. 5097, pp. 147-149.

[63] Tivey, Maurice A., H. Paul Johnson, Corinne Fleutelot, Stefan Hussenoeder, Roisin Lawrence, Cheryl Waters, Beecher Wooding. October 1, 1998. "Direct measurement of magnetic reversal polarity boundaries in a cross-section of oceanic crust." *Geophysical Research Letters,* Vol. 25, No. 19, pp. 3631-3634.

[64] Hall, J.M., P.T. Robinson. 1979. "Deep crustal drilling in the North Atlantic Ocean." *Science*, Vol. 204, pp. 573-586.

[65] Agocs, W.B., Meyerhoff, A.A., and Kis, K. 1992. "Reykjanes Ridge: quantitative deter minations from magnetic anomalies." In Chatterjee, S. and Hotton, N., III, eds., *New Concepts in Global Tectonics*, Lubbock, TX: Texas Tech University Press, pp. 228.

[66] Storetvedt, Karsten. 2010. "World Magnetic Anomaly Map and Global Tectonics." *New Concepts in Global Tectonics Newsletter*, no. 57, Dec. 2010. p. 28.

[67] Agocs, 1992. p. 225.

[68] Liebke, U., E. Appel, U. Neumann, L. Ding. 2012. "Dual polarity directions in basaltic-andesitic dykes—reversal record or self-reversed magnetization?" *Geophysical Journal International*, Vol. 190, pp. 887-899.

[69] Rea, David K., and John M. Dixon. 1983. "Late Cretaceous and Tectonic Evolution of the North Pacific Ocean." *Earth and Planetary Science Letters*, 65. pp. 156-161.

[70] Agocs, 1992. pp. 228-229.

[71] Smith, Norm, and Jane Smith. 1993. "An alternative explanation of oceanic magnetic anomaly patterns." *Origins*, Vol. 20, No. 1, pp. 6-21.

[72] Austin, S.A. 1996. "Excess argon within mineral concentrates from the new dacite lava dome at Mount St Helens volcano." *Journal of Creation*, 10(3):335–343.

[73] Kasahara, Junzo. 2002. "Tides, Earthquakes, and Volcanoes." *Science*, Vol. 297. July 19, 2002. pp. 348-349.

[74] Witze, Alexandra. 2013. "Quakes in Slo-Mo." *Science News*. Vol. 183. pp. 28.
Rogers, Garry, and Herb Dragert. 2003. "Episodic Tremor and Slip in the Cascadia Subduction Zone: The Chatter of Silent Slip."

INDEX

www.ingramcontent.com/pod-product-compliance
Lightning Source LLC
Chambersburg PA
CBHW042337040426
42447CB00017B/3466